T0303909

Fundamentals of MRI
An Interactive Learning Approach

Series in Medical Physics and Biomedical Engineering

Series Editors: John G Webster, E Russell Ritenour, Slavik Tabakov,
and Kwan-Hoong Ng

Other recent books in the series:

Handbook of Optical Sensing of Glucose in Biological Fluids and Tissues
Valery V. Tuchin (Ed)

A Introduction to Radiation Protection in Medicine
Jamie V. Trapp and Tomas Kron (Eds)

A Practical Approach to Medical Image Processing
Elizabeth Berry

Biomolecular Action of Ionizing Radiation
Shirley Lehnert

An Introduction to Rehabilitation Engineering
R A Cooper, H Ohnabe, and D A Hobson

The Physics of Modern Brachytherapy for Oncology
D Baltas, N Zamboglou, and L Sakelliou

Electrical Impedance Tomography
D Holder (Ed)

Contemporary IMRT
S Webb

The Physical Measurement of Bone
C M Langton and C F Njeh (Eds)

**Therapeutic Applications of Monte Carlo Calculations
in Nuclear Medicine**
H Zaidi and G Sgouros (Eds)

Minimally Invasive Medical Technology
J G Webster (Ed)

Intensity-Modulated Radiation Therapy
S Webb

Physics for Diagnostic Radiology, *Second Edition*
P Dendy and B Heaton

Achieving Quality in Brachytherapy
B R Thomadsen

Medical Physics and Biomedical Engineering
B H Brown, R H Smallwood, D C Barber, P V Lawford, and D R Hose

Series in Medical Physics and Biomedical Engineering

Fundamentals of MRI
An Interactive Learning Approach

Elizabeth Berry
Elizabeth Berry Ltd
Leeds, UK

Andrew Bulpitt
University of Leeds
School of Computing, Leeds, UK

CRC Press
Taylor & Francis Group
Boca Raton London New York

CRC Press is an imprint of the
Taylor & Francis Group, an **informa** business

A TAYLOR & FRANCIS BOOK

CRC Press
Taylor & Francis Group
6000 Broken Sound Parkway NW, Suite 300
Boca Raton, FL 33487-2742

© 2009 by Taylor & Francis Group, LLC
CRC Press is an imprint of Taylor & Francis Group, an Informa business

Library of Congress Cataloging-in-Publication Data

Berry, Elizabeth, 1961-
    Fundamentals of MRI : an interactive learning approach / Elizabeth Berry, Andrew J. Bulpitt.
        p. cm. -- (Series in medical physics and biomedical engineering)
    Includes bibliographical references and index.
    ISBN 978-1-58488-901-4 (hardcover : alk. paper)
    1. Magnetic resonance imaging--Programmed instruction. I. Bulpitt, Andrew J. II. Title. III. Series.

QC762.6.M34B47 2008
616.07'548--dc22   2008044195

Visit the Taylor & Francis Web site at
http://www.taylorandfrancis.com

and the CRC Press Web site at
http://www.crcpress.com

Dedication

To David, Jane, and Douglas

To Jayne, Megan, Jemma, and Lucy

Contents

Preface

The physics of magnetic resonance imaging is a little harder to understand than the physics of other medical imaging modalities, but it is extremely rewarding to develop an understanding because it makes future study and work in the field much easier. *Fundamentals of MRI: An Interactive Learning Approach* has been written to help readers gain confidence in the basic concepts that underpin the field. The book and CD include more learning material than is usual; there are questions integrated throughout the text and interactive exercises that make use of specially written computer programs. Answers to all the activities are included, so the materials are suitable for those undertaking self-directed study as well as those undertaking academic study in medical imaging and related fields in a more formal learning environment, with or without a clinical component.

Elizabeth Berry
Andrew Bulpitt

Acknowledgments

I gratefully acknowledge the many colleagues from whom I've learned about MRI, especially John Ridgway.

E. B.

I would like to thank my colleagues for the advice and solutions offered to resolve my Java queries.

A. B.

We are grateful to all those who gave permission for the reproduction of their images or data.

About the Authors

Elizabeth Berry, Ph.D., was born in 1961. She received a B.Sc. degree in physics from the University of Hull in 1982 and a Ph.D. in applied optics from Imperial College London in 1986. She held medical physicist posts in Inverness and Bristol until 1993. From 1993 to 2005, she was lecturer and then senior lecturer in medical imaging at the University of Leeds. Since 2005, she has been the director of Elizabeth Berry Ltd., and is currently an associate lecturer with the Open University. Dr. Berry is a chartered scientist, fellow of the Institute of Physics, and fellow of the Institute of Physics and Engineering in Medicine.

Andrew Bulpitt, D.Phil., was born in 1969. He received a B.Eng. degree in electronic engineering from the University of Liverpool in 1990 and a D.Phil. in electronics from the University of York in 1994. From 1994 to 2000, he was a research fellow in medical imaging in the School of Computing at the University of Leeds. He was appointed a lecturer in computing in 2000 and is currently a senior lecturer in the school. Dr. Bulpitt is a member of the Institute of Electronic Engineers.

1

Introduction

Magnetic resonance imaging (MRI) is probably the most flexible of the medical imaging modalities. Images in MRI can be acquired using many different acquisition techniques, known as imaging pulse sequences, and a whole range of images with differing contrast properties can be achieved. A tissue can be made to give a signal higher or lower than another tissue, or even to give no signal at all. All these possibilities are a direct consequence of the physical principles that underpin the technique, which are fundamentally different from those of other medical imaging methods. You may already be familiar with x-ray imaging, in which images are formed by detecting the x-rays after they have been attenuated by tissue. X-rays are electromagnetic radiation with a high frequency. Electromagnetic radiation is involved in MRI too, but in a very different way. In MRI the radiation has a much lower frequency, and is not attenuated by tissue and then detected. Instead, the radiation is one of the components of the process of nuclear magnetic resonance, which also requires an interaction between atomic nuclei and a strong magnetic field. For MRI, the nuclei concerned are hydrogen nuclei, which are abundant in the human body. Measuring the signals that the hydrogen nuclei emit gives information about the tissue in which they are situated, and much of the practice of MRI is concerned with ensuring that signals from hydrogen nuclei in an environment of interest can be clearly distinguished. To engage with all the different ways in which an MR image can be acquired requires an understanding of the basic principles and concepts. Such an understanding is hugely valuable and provides a solid foundation because the same principles form the basis of every imaging pulse sequence or new development in the field.

1.1 The Fundamentals of MRI

This is a book about the fundamentals of magnetic resonance imaging, and it is intended for readers just beginning to learn about MRI. Our aim is to help the reader to gain confidence with the core concepts before he or she moves on to further study or practical training.

1.2 An Interactive Learning Approach

The contents of the book and the programs on the CD have been designed to promote active learning. *Active learning* is the term used to describe learning by thinking, doing, and reflecting, and for many learners is a more productive approach than more passive methods. The materials are designed to support different learning styles and have been written with the knowledge that some readers will find equations informative, while others may prefer a wordy explanation of relationships between parameters.

In this book, the interactive learning approach means that not only is factual information presented, but there are also questions and exercises integrated with the text, to help reinforce the concepts studied. Answers follow immediately after the questions because the questions are part of the learning process rather than a testing tool. The 10 interactive programs on the CD are an integral part of the book. Four examples of the user interfaces to the programs are shown in Figure 1.1. The programs not only allow the concepts introduced in the printed material to be clearly demonstrated and further developed, but also provide the reader with the opportunity to engage in the learning process through guided exercises and by exploring the concepts in his or her own way.

It is recommended that you install the programs before starting to work through the book, so that you can access them easily when they are needed. If you are using a Windows®* operating system, this process is described in the next section. If you are using a different operating system, read Section 1.4 instead.

1.3 Using the Programs from Windows® Operating Systems

1.3.1 Program Setup Options

The software provided on the CD is written in Java®† to enable it to be used on different operating systems. The programs therefore require a compatible version of the Java Runtime Environment to be used. The setup program for Windows determines if Java is already installed on your computer and whether or not the version you have is compatible with the programs provided on the CD. If not, it will give you the option to install the version of Java supplied on the CD.

If you do not have the rights to install software on your computer, perhaps because you are using a shared computer or one supplied by your

---

* Registered trademark of Microsoft Corporation, Redmond, Washington, United States.
† Java and all Java-based marks are trademarks or registered trademarks of Sun Microsystems, Inc., Santa Clara, California, in the United States and other countries.

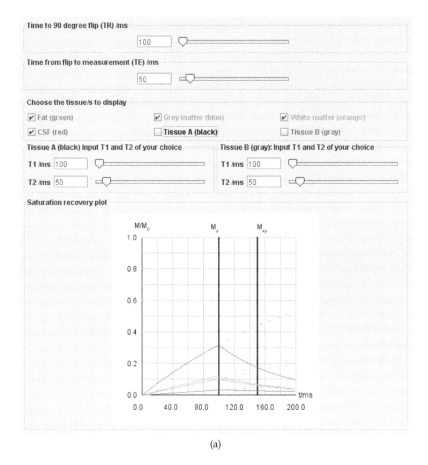

(a)

FIGURE 1.1

Examples of the user interfaces for the interactive programs that are integrated with the material in this book. (a) The *Saturation Recovery* program (Chapters 4 and 9), (b) the *Slice Selection* program (Chapter 5), (c) the *Phase Encoding Demonstrator* (Chapter 7), and (d) the *Flow Phenomena Demonstrator* (Chapter 10).

employer, or if you simply do not wish to install any software, this is not a problem. The programs can be run directly from the CD, as indicated in the next section.

1.3.2 Running Programs Directly from the CD

If you do not wish to run the installation program, or if the automatic setup fails, you can run the programs directly from CD:

- Insert the CD into the CD-ROM/DVD drive.
- Open the "Fundamentals of MRI (X:)" folder in *My Computer* or Windows Explorer.

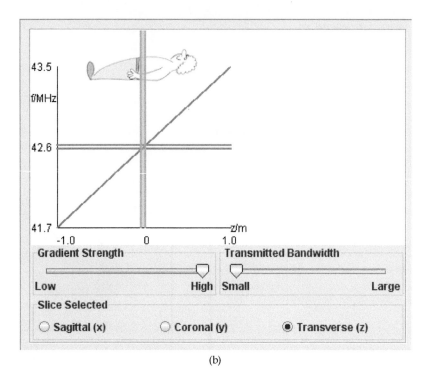

(b)

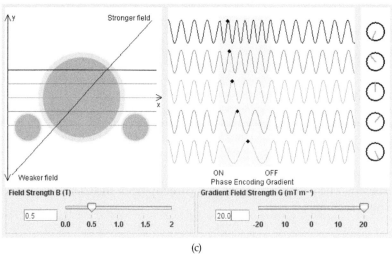

(c)

FIGURE 1.1   *(continued)*

- Locate and double-click on RunFromCD.bat.
- Select the program you require from the menu under the *Run from CD* tab (Figure 1.2).

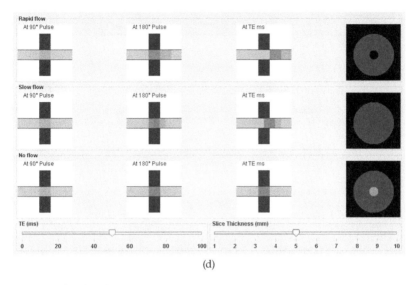

(d)

FIGURE 1.1 (*continued*)

1.3.3 Running the Setup Program

The setup program (which is called `FundamentalsOfMri.bat`) will determine if Java is already installed on your computer and, if required, will give you the option to install the version of Java supplied on the CD. Note that once the setup program has determined which version of Java to use, it will use this version whenever you run the programs in the future.

- Insert the CD into the CD-ROM/DVD drive:
  a. If the CD AutoRun feature is enabled, the installation will test to see if Java exists on your computer, and if it is already installed, the window shown in Figure 1.2 will appear on your monitor.
  b. If the CD AutoRun feature is disabled:
     - Double-click to open the CD entitled "Fundamentals of MRI (X:)" in *My Computer* or Windows Explorer.
     - If the window shown in Figure 1.2 does not open, locate and double-click on `FundamentalsOfMri.bat` on the CD.
- If a compatible version of Java was found not to exist on your computer, then the Java Runtime Environment setup program will be launched. If you wish to install Java, follow the instructions on the screen.
- Do not worry if the Java Runtime installation is unsuccessful, or if you do not wish to install Java. You will still be able to access

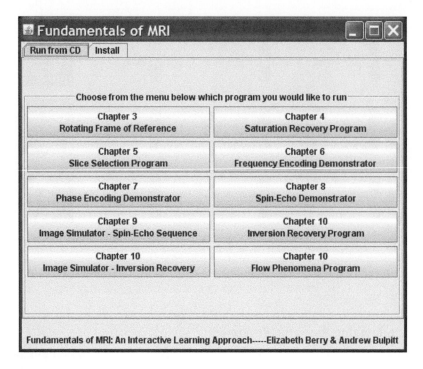

FIGURE 1.2

The user interface for the *Fundamentals of MRI* programs. If you choose to install the programs to your computer, you will see this window only at the time of installation. If you
choose to run the programs from CD, then you will see this window every time you wish
to run a program.

the programs directly from the CD, as indicated in Section 1.3.2,
using the file RunFromCD.bat.

1.3.4 Installing the Programs

The setup program also provides you with the option to install all the programs on your computer. This means that you will not have to use the CD
every time, but it has the additional benefit that the programs will be automatically updated from the book Web site at www.fundamentalsofmri.
com. Downloads of the latest versions of the programs will occur whenever an update is detected, so installing the programs will allow us to
provide you with updates of programs automatically if required.

- Select the tab labeled *Install*. The window shown in Figure 1.3
  will appear.
- Click on the *Install All Programs* button.

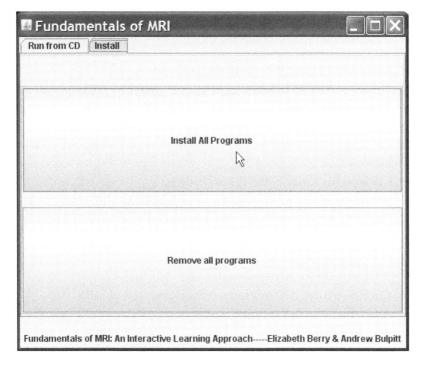

FIGURE 1.3
The installation interface for the *Fundamentals of MRI* programs. Installing the programs means that downloads of the latest versions of the programs will occur automatically.

- The programs will be installed to your computer and short-cuts to the programs will be added from the start menu, under *Fundamentals of MRI*.
- When the installation is complete, a confirmation message will appear, and you can close the program using the cross in the top-right-hand corner.

1.3.5 Running the Installed Programs

- The CD is not required.
- Click on *Start | All Programs | Fundamentals of MRI*
- Select the program that you require from the list.

Note that the first time the programs are run it may be necessary to set the computer's firewall so that Java can operate and the book website be accessed for updates.

1.3.6 Uninstalling the Programs

It is recommended that you use the CD:

- Insert the CD into the CD-ROM/DVD drive.
- Open the "Fundamentals of MRI (X:)" folder in *My Computer* or Windows Explorer.
- Locate and double-click on FundamentalsOfMRI.bat.
- Select the tab labeled *Install.* The window shown in Figure 1.3 will appear.
- Click on the *Remove all programs* button.
- When the removal of programs is complete a confirmation message will appear, and you can close the program using the cross in the top right-hand corner.

If you do not have the CD:

- Click on Start | All Programs | Fundamentals of MRI.
- Select the *Uninstall Program* option from the list, and wait—there may be a delay.
- If a security warning appears, click on the *Run* option.
- Click on the *Remove all programs* button.
- When the removal of programs is complete a confirmation message will appear, and you can close the program using the cross in the top right-hand corner.

1.4 Non-Windows Operating Systems

The software provided on the CD is written in Java to enable it to be used on different operating systems. The programs can be used from non-Windows operating systems if Java is already installed on your computer.

- Insert the CD into the CD-ROM/DVD drive.
- Run the file OtherOS.jar. In some cases, double-clicking on the file in a window will work; otherwise, enter java -jar OtherOS. jar at the command line.
- The window shown in Figure 1.2 will appear.
- Select the program you require from the menu under the *Run from CD* tab.

1.5 Structure of the Book

The first two chapters form the introduction to the book, with information in this chapter about the computer programs, and background information that is needed for understanding material presented later in the book provided in Chapter 2. Background mathematics, physics, and digital imaging are all included, and there is also a table of key clinical imaging terms. Some readers will be happy to skim Chapter 2, returning to it as a reference resource when required. Others may prefer to work through the whole chapter, or specific parts, before proceeding to the rest of the book. For the latter group, the first 50 multiple-choice questions in Chapter 11 cover the same ground and will be useful for revision.

The main material is contained in Chapters 3 to 10 and the associated *Fundamentals of MRI* programs. Activities involving the programs form a significant part of each chapter, which is one reason it is recommended that the programs be installed on your computer, where they will be easily accessible when needed. In every chapter, questions and exercises are integrated within the text. The questions form part of the learning process, and because they are not there for assessment, the answer is provided immediately after each question. Readers will quickly get used to ignoring the text immediately after each question until they are ready to read it. Most of the chapters introduce new concepts and give many opportunities to work toward a satisfactory understanding. In Chapter 10, although new ideas are introduced, they can be explained by applying the concepts covered in the preceding chapters. This will be the case for many other topics in the field when they are encountered in the future.

The final chapter of the book contains multiple-choice questions that are suitable for revision and reflection. The 150 questions are organized in the same order as the material appears in the book, and in such a way that the reader can select a subset of questions either to cover all of the material or to concentrate on specific areas. Answers, and some further feedback, appear separately at the end of the chapter.

2

Mathematics, Physics, and Imaging for MRI

This chapter contains a range of background information that is assumed in the later chapters of the book. Few people, however, will want to work through the whole chapter in detail before starting to read about magnetic resonance imaging (MRI) in Chapter 3. The best approach is to skim through this chapter on first reading the book, so that you have a rough idea of what it covers. Then refer back to it when required. Four subjects are covered. The first and longest section concerns mathematics. Concepts including vectors, the exponential function, and Fourier analysis are included here. There is also a table that highlights where in the book the mathematical concepts are applied. The second subject is physics. Just a few areas are included to underpin the discussion of the basic principles of MRI in Chapter 3. Concepts associated with digital images (such as the pixel and signal-to-noise ratio) are outlined in the third section. The chapter closes with definitions of key clinical imaging terms. Multiple-choice questions on all of this material are included in Chapter 11.

2.1 Learning Outcomes

When you have worked through this chapter you should have sufficient understanding of the aspects of mathematics, physics, imaging, and clinical imaging terminology covered here to follow the material presented in the rest of the book.

2.2 Mathematics for MRI

2.2.1 Trigonometric Functions

2.2.1.1 Sine, Cosine, and Tangent

The sine, cosine, and tangent of an angle can be defined in a right-angled triangle (Figure 2.1). The sides have lengths a, b, and c. The side (c) opposite

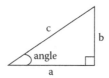

FIGURE 2.1
A right-angled triangle, used to define the relationships for sine, cosine, and tangent.

the right angle is called the *hypotenuse*. The angle *angle* can be expressed in terms of the side lengths in three ways:

$$\sin\left(angle\right) = \frac{b}{c} \tag{2.1}$$

$$\cos\left(angle\right) = \frac{a}{c} \tag{2.2}$$

$$\tan\left(angle\right) = \frac{b}{a} = \frac{\sin\left(angle\right)}{\cos\left(angle\right)} \tag{2.3}$$

Some people find it helpful to use a mnemonic for remembering these relationships. To use with a mnemonic, the sides of the triangle need to be renamed as in Figure 2.2, and the equations arranged as

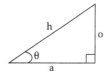

$$t = \frac{o}{a}, \, s = \frac{o}{h}, \, c = \frac{a}{h}$$

where $t = \tan\theta$, $s = \sin\theta$, and $c = \cos\theta$, and o, a, and h are as shown in Figure 2.2. These letters (toasohcah) can then be used as the first letters of the words in a nine-word sentence that you find easy to remember.

FIGURE 2.2
Right-angled triangle, with the sides labeled to help with remembering the expressions for sine, cosine, and tangent of the angle θ. The side adjacent to the angle θ is a, the side opposite the angle θ is o, and the hypotenuse (the side opposite the right angle) is h.

2.2.1.2 Inverse Sine, Cosine, and Tangent

The expressions for sine, cosine, and tangent can be inverted. For example, the angle whose tangent is b/a (Figure 2.1) can be expressed in two different ways (Equations 2.4 and 2.5), which mean the same thing:

$$angle = \tan^{-1}\left(\frac{b}{a}\right) \tag{2.4}$$

$$angle = \arctan\left(\frac{b}{a}\right) \tag{2.5}$$

Similarly,

$$angle = \sin^{-1}\left(\frac{b}{c}\right) \tag{2.6}$$

$$angle = \cos^{-1}\left(\frac{a}{c}\right) \tag{2.7}$$

2.2.1.3 Relationships between the Trigonometric Functions

Note that the square of a sine, tangent, or cosine is usually written with the square notation immediately following the symbol sin, cos, or tan. For example, the square of the cosine of *angle* is written $\cos^2(angle)$ rather than the more unwieldy $[\cos(angle)]^2$. The following relationship is sometimes needed:

$$\sin^2(angle) + \cos^2(angle) = 1 \tag{2.8}$$

2.2.1.4 Degrees and Radians

You will be familiar with degrees, which are units used to express the size of angles. For example, a right angle is 90°, and 360° corresponds with a full rotation. Radians are an alternative unit for measuring angles, and they allow angles to be expressed as fractions or multiples of the constant π. Some useful angles are shown in Table 2.1.

TABLE 2.1

Commonly Used Angles

Degrees	Radians
0	0
90	$\pi/2$
180	π
270	$3\pi/2$
360	2π

Note: The angle $\pi/2$ is usually read as "pi by two."

2.2.1.5 Phase Differences

The sine and cosine of an angle change in a predictable way as the size of the angle is

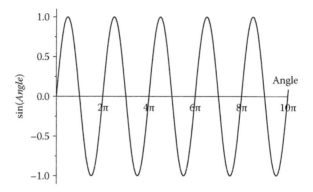

FIGURE 2.3

The change in the value of the function sin(*Angle*) with *Angle*. The value of *Angle* increases along the horizontal axis from 0 to 10π radians. The function has values between –1 and +1. Note how the function crosses the horizontal axis (where the function has the value 0) at 0, π, 2π, 3π, 4π.... The wavelength of sin(*Angle*) is 2π.

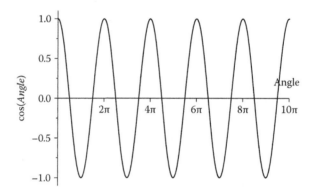

FIGURE 2.4

The change in the value of the function cos(*Angle*) with *Angle*. The value of *Angle* increases along the horizontal axis from 0 to 10π radians. Like sin(*Angle*), the function has values between –1 and +1, and a wavelength of 2π. However, cos(*Angle*) is shifted with respect to sin(*Angle*); cos(*Angle*) has the value +1 when *Angle* = 0, but sin(*Angle*) has the value 0 at *Angle* = 0.

changed. The function sin(*Angle*) is shown in Figure 2.3, and cos(*Angle*) in Figure 2.4. The two functions look rather similar, and both have a wavelength of 2π. However, they are shifted along the *Angle* axis with respect to one another. If both functions are plotted on the same graph, this shift can be seen more clearly (Figure 2.5). In Figure 2.5, only one cycle of the wave is plotted. We see that sin(*Angle*) is shifted to the right of cos(*Angle*) by π/2 radians (that is, by 90°). This shift is called a *phase shift* or *phase difference*. The functions sin(*Angle*) and cos(*Angle*) are identical except for a phase difference of π/2.

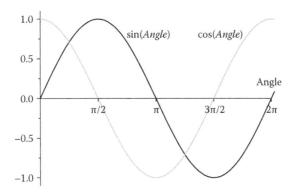

FIGURE 2.5
The relationship between the functions sin(*Angle*) and cos(*Angle*).

2.2.2 Vectors

2.2.2.1 Definition of a Vector

Vectors are used to define quantities that have both a size (magnitude or amplitude) and a direction. For example, velocity is a vector, but speed is not. Vectors can be plotted using the familiar axes of rectangular (Cartesian) coordinates, but a compact way to represent the two quantities associated with vectors is to use polar coordinates. The two variables in *polar coordinates* are amplitude and angle. A vector in polar coordinates is illustrated in Figure 2.6.

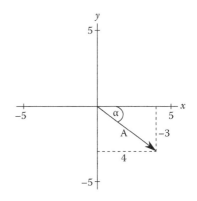

FIGURE 2.6
A vector illustrated using rectangular and polar coordinates. The amplitude, A, of the vector is 5 and the angle α is –37°.

The amplitude *A* of the vector is given by

$$A = \sqrt{x^2 + y^2} \tag{2.9}$$

The angle of the vector α is

$$\alpha = \tan^{-1}\left(\frac{y}{x}\right) \tag{2.10}$$

Equation 2.10 gives the angle between the vector and the *x* axis. The *phase* of a vector is an angle between 0 and 360°, measured in an anticlockwise direction from the positive *x* axis. For the vector in Figure 2.6, the phase is (360 – 37)° = 323°.

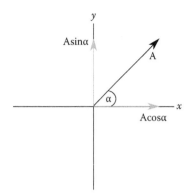

FIGURE 2.7
The vector is broken down, or resolved, into two component vectors (gray) at right angles to each other.

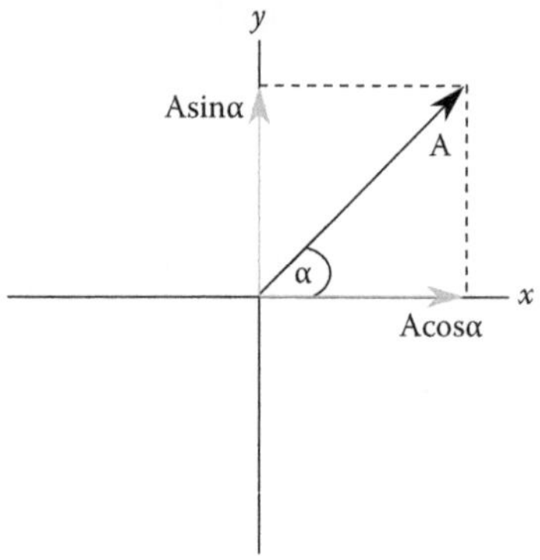

FIGURE 2.8
How to drop perpendiculars (dashed lines) to resolve vectors.

2.2.2.2 Resolving Vectors

A vector can be broken down into component vectors at right angles to each other. These are known as *orthogonal components*, and the process is known as *resolving* a vector (Figure 2.7). The size of the x component is given by $A\cos\alpha$, and the size of the y component by $A\sin\alpha$ (using the basic trigonometry covered in Section 2.2.1). The components can be indicated in a sketch by dropping perpendiculars from the vector to the axes (Figure 2.8). In the example of Figure 2.6, the x component has the value 4 and the y component is −3.

Resolving vectors into two orthogonal components is very useful because the two components can be treated independently. This independence arises because the magnitude of the component of a vector in a direction at 90° to its own line of action is always equal to zero.

Question 2.1

(a) In Figure 2.9a, what is the y component of the vector shown?
(b) In Figure 2.9b, what are the x and y components of the vector shown?
(c) In Figure 2.9c, what is the x component of the vector shown?
(d) In Figure 2.9d, if the y component of the vector is 4, what is the phase angle α of the vector?
(e) Sketch Figure 2.9e and add the x and y components of the vector.
(f) Sketch Figure 2.9f and add the x and y components of the vector.
(g) Sketch Figure 2.9g and add the x and y components of the vector.
(h) Sketch Figure 2.9h and add the x and y components of the vector.

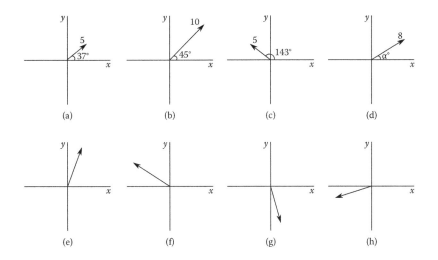

FIGURE 2.9
(a–h) Diagrams for Question 2.1.

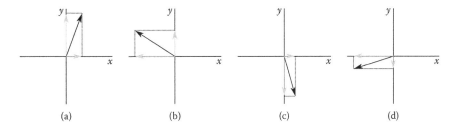

FIGURE 2.10
(a–d) Feedback on Question 2.1, parts (e) to (h).

Answer

The answers are (a) y = 3, (b) x = 7.07, y = 7.07, (c) x = −4, (d) α = 30°. The completed sketches for parts (e) to (h) are shown in Figure 2.10.

2.2.2.3 Adding Vectors

When vectors are added it is necessary to take into account both their magnitude and their direction. For example, two vectors might have equal magnitude, but act in opposite directions. In this case, the result of adding the two vectors is zero.

The easiest way to add vectors, particularly if there are lots of them, is first to resolve each vector into its x and y components, then to add up each component separately. For example, consider the five vectors in

TABLE 2.2

The Five Vectors Shown in Figure 2.11

Vector	x Component	y Component
a	4	3
b	0	5
c	−4	−3
d	5	0
e	−3	4
Sum	2	9

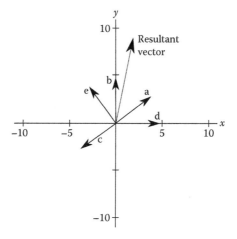

FIGURE 2.11

Vector addition using components. The magnitude of the resultant vector for the vectors listed in Table 2.2 is $\sqrt{(2^2 + 9^2)} = 9.2$.

Table 2.2 and Figure 2.11, which all have the magnitude 5 but have different directions.

Question 2.2

Calculate the values for sums and vector magnitudes that should appear in the empty cells marked with an asterisk in Table 2.3. Draw a sketch showing each of the vectors listed in the table and their sum.

Answer

The vector magnitudes in the right-hand column of the table all have the value 13. The sum of the x components is −5, and the sum of the y components is 22. The vector magnitude is 22.56. A sketch of the vectors is in Figure 2.12.

TABLE 2.3

Table for Question 2.2

Vector	x Component	y Component	Vector Magnitude
a	5	12	*
b	12	5	*
c	−5	−12	*
d	−5	12	*
e	−12	5	*
Sum	*	*	—
		Resultant vector magnitude	*

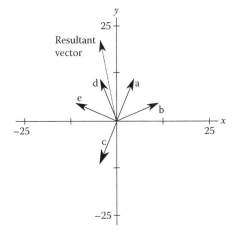

FIGURE 2.12
Feedback for Question 2.2.

Question 2.3

Calculate the values that should appear in the empty cells marked with an asterisk in Table 2.4, and draw a sketch showing each of the vectors and their sum.

Answer

The x component of vector a is 3. The vector magnitude of vector b is 5. The y component of vector c is 24. The vector magnitude of vector d is 15. The sum of the x components is 5, and the sum of the y components is 37. The vector magnitude is 37.3. A sketch of the vectors is shown in Figure 2.13.

TABLE 2.4

Table for Question 2.3

Vector	x Component	y Component	Vector Magnitude
a	*	4	5
b	4	–3	*
c	7	*	25
d	–9	12	*
Sum	*	*	—
		Resultant vector magnitude	*

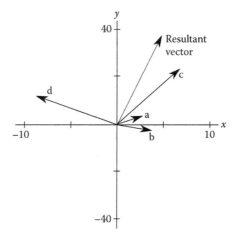

FIGURE 2.13
Feedback for Question 2.3.

Question 2.4

Calculate the values that should appear in the empty cells marked with an asterisk in Table 2.5, and draw a sketch showing each of the vectors and their sum.

Answer

The sum of the x components is 0, the sum of the y components is 0, and the resultant vector magnitude is 0. A sketch of the vectors is shown in Figure 2.14. This is a special case of the situation where there is a large number of randomly oriented vectors, which cancel each other out and have a resultant of 0.

TABLE 2.5

Table for Question 2.4

Vector	x Component	y Component	Vector Magnitude
a	0	1	1
b	0	−1	1
c	1	0	1
d	−1	0	1
Sum	*	*	—
		Resultant vector magnitude	*

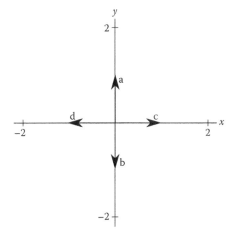

FIGURE 2.14
Feedback for Question 2.4.

2.2.3 Three-Dimensional Vectors

In MRI we will work with three-dimensional vectors (Figure 2.15). Usually the three-dimensional vector is described in terms of two components. The first is the component in the z direction (longitudinal), and the second is a single component in the xy plane (transverse). Just like any other two-dimensional vector, the xy component could itself be resolved into separate x and y components. These separate components in the xy plane are sometimes called the *real* and *imaginary* components, or the *in-phase* and *in-quadrature* components, of the MR signal.

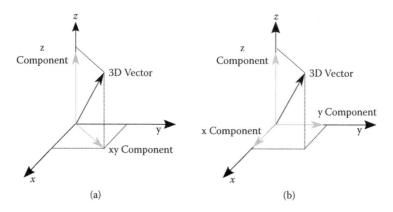

FIGURE 2.15
(a) A three-dimensional vector shown with two resolved components. (b) The same three-dimensional vector resolved into three components.

2.2.4 Periodic Signals

Periodic signals, or waveforms, are signals made up of a repeating pattern. The sine and cosine functions, which were introduced in Section 2.2.1, are examples of periodic signals, but there are many others. Examples are shown in Figure 2.16. A *cycle* is the part of the waveform that is repeated, and this term applies whether the variation is taking place in time or across a distance. The length of one cycle is the *wavelength*, λ. For a temporal variation, the time for one cycle is called the *period*.

Linear frequency, f, is measured in cycles per second, a unit which is also named the hertz (Hz). Linear frequency is the number of complete cycles in one second.

The rate of variation can also be expressed as an *angular frequency, ω*. Angular frequency is measured in radians per second. The relationship between linear and angular frequency is

$$\omega = 2\pi f \qquad\qquad (2.11)$$

2.2.5 Phase Diagrams

Periodic functions can be plotted using an angular coordinate system in order to illustrate the phase of the function, and phase differences between functions, more clearly. These diagrams are called phase diagrams. In a phase diagram an arrow, known as a *phasor*, rotates around a fixed point through an angle (from 0 to 2π) measured in an anticlockwise direction from the positive horizontal axis. The phase diagram for the sine function is shown in Figure 2.17. The functions sine and cosine have a phase difference of π/2 radians, so if plotted together on a phase diagram they will always appear separated by an angle of π/2 radians (90°).

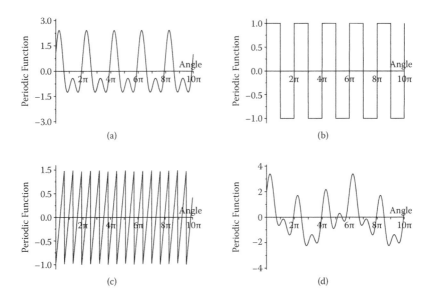

FIGURE 2.16
Examples of periodic signals. The change in the value of four different periodic functions with the value of *Angle*. The value of *Angle* increases along the horizontal axis from 0 to 10π radians. The length of the repeated cycle is (a) 2π, (b) 2π, (c) 2, and (d) 6π.

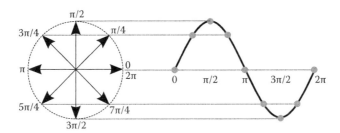

FIGURE 2.17
The relationship between the phase diagram of a sine wave and the conventional plot. At the left a phasor is shown at various positions during its anticlockwise rotation through 2π radians (360°). The magnitude of the phasor in the vertical direction corresponds with the value of the sine wave at that angle, so the tip of the phasor traces out a sine wave (right) when plotted against angle.

2.2.6 Sampling

A continuous function, which is a function defined by an equation, can have a value calculated for it at every point on the curve. In practice, when a signal is detected, measurements of the signal are made only at discrete intervals. In other words, sampling has taken place, meaning that values are known only at certain points. To reproduce the underlying signal faithfully,

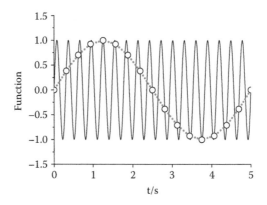

FIGURE 2.18
The periodic signal shown with the solid black line has a frequency of 3.4 Hz. To reproduce the frequency of the signal correctly, a sampling frequency of a least twice the signal frequency would be required. In this case, the signal has been sampled at a lower frequency (3.2 Hz), at the locations shown by the circles. This undersampling results in an aliased signal (shown gray and dotted) with a frequency of 0.2 Hz.

sampling needs to take place at sufficiently close spacing, with more closely spaced samples being necessary to reproduce high-frequency variations. Quantitatively, it is known from the Shannon sampling theorem that the sampling frequency must be double the highest frequency that it is desired to detect in the signal. The principle is illustrated in Figure 2.18. If the sampling frequency is too small compared with the frequency of the signal, then the signal will be *undersampled*, and the resulting measurement is described as being *aliased*. When aliasing has occurred, a sampled high-frequency signal appears to have a lower frequency than the original signal.

2.2.7 Fourier Series and Fourier Analysis

2.2.7.1 Fourier Series

Any function, $f(x)$, may be expressed as the sum of periodic functions, providing that $f(x)$ is itself a periodic function and is defined over a fixed interval. This sum of terms is known as a *Fourier series*, and it consists of a fundamental frequency plus a series of harmonics at multiples of that frequency. For example, the Fourier series for the triangle function shown in Figure 2.19a is given by Equation 2.12:

$$f(x) = \frac{1}{4} + \frac{4}{\pi^2}\cos(x) + \frac{8}{(2\pi)^2}\cos(2x) + \frac{4}{(3\pi)^2}\cos(3x) + \frac{4}{(5\pi)^2}\cos(5x) + \dots \quad (2.12)$$

Figure 2.19b,c,d,e,f show the series including only 1, 2, 4, 8, and 16 cosine terms in the series. It can be seen that the later terms in the series, which

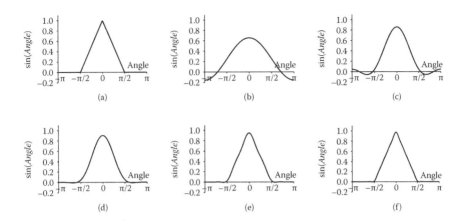

FIGURE 2.19
(a) Triangle function built using the first 50 terms of the Fourier series in Equation 2.12. The results using (b) 1, (c) 2, (d) 4, (e) 8, and (f) 16 cosine terms of the same Fourier series. Note how the later, higher-frequency terms in the series contribute to sharpness of the corners of the shape.

are the higher-frequency terms, are needed to form the sharp features in the shape.

2.2.7.2 Fourier Analysis

Fourier analysis is the generalization of the Fourier series principle, without the requirement for a periodic function in a fixed interval. Fourier analysis may be applied to functions that have no particular periodicity and can be used over an infinite interval. In the same way as the Fourier series can be used, Fourier analysis is a way of expressing a complex function as a series of terms, each term having a single frequency. It is used as a method of extracting information about the composition of a measured signal.

For MRI, the important points about Fourier analysis are:

- Any function can be expressed as a series of sine and cosine terms, with each term at a single frequency. If the result is plotted in a graph of amplitude versus frequency, then the graph is called a spectrum. The spectrum is a visual indication of the amount of each frequency present in the function.
- The higher-frequency terms are necessary to reproduce fine detail in the function.
- A one-dimensional Fourier transform is performed on a one-dimensional series of numbers. A two-dimensional Fourier transform requires a two-dimensional array of numbers.

- Fourier transformation is performed on measured signals in MRI. The signal replaces the continuous function that we used to introduce the concept. So the result of the Fourier transform shows how much of each frequency is present in the signal.
- The fast Fourier transform (FFT) is a computational method for performing Fourier transformation.

2.2.8 Exponential Decay

The exponential function is a mathematical function that has interesting properties. A general equation representing exponential decay can be written in two ways:

$$A = A_0 e^{-bx} \text{ or } A = A_0 \exp\left(-bx\right) \tag{2.13}$$

The two ways of writing the exponential function mean exactly the same thing. In the expressions A_0 and b are both constants, and x is a variable. The minus sign associated with the power of the exponential ensures that the expression describes decay. e is a mathematical constant that always has the value $e = 2.718$. A plot of A against x is shown in Figure 2.20. Notice that when $x = 0$, the result is A_0. This is because any number raised to the power 0 is 1. In Figure 2.20, the change in x needed to reduce A from A_0 to $A_0/2$ is indicated. The change in x to further reduce A from $A_0/2$ to $A_0/4$ is also shown: it is exactly the same change in x. It turns out that this property can be generalized, and that the reduction of A by any factor, not just by 2, will always be associated with an increase in x that is fixed for that factor.

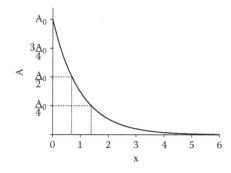

FIGURE 2.20
The form of the exponential decay function of Equation 2.13. $b = 1$ in this example.

2.2.8.1 Time Constant

The exponential decays encountered in MRI usually have time as the variable (Figure 2.21), and the general form of the exponential decay is written as $\exp(-t/\tau)$. We have replaced the variable x in Equation 2.13 by t, and the constant b by $1/\tau$. τ is called the time constant of the decay, measured in the same units as t. The time constant is defined as the time taken to get to a point on the curve at which the value is e^{-1}, in other words, the point where $t = \tau$. Since e has the value 2.718, this corresponds to the point where 0.37 of the original signal remains. The property of the exponential decay of equal time for the same fractional reduction means that the value of the decay curve will fall by 0.37 for every increment in time of size τ. Figure 2.22 shows how the shape of the exponential decay curve is

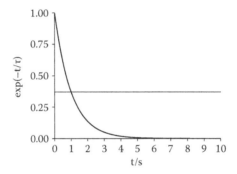

FIGURE 2.21
The exponential decay of Equation 2.13 plotted as a function of time and with $b = 1/\tau$, where τ is the time constant. The time constant is measured in units of time, and in this example $\tau = 1$ s. When $t = \tau$, the value of the function is 0.37, which is indicated by the horizontal line.

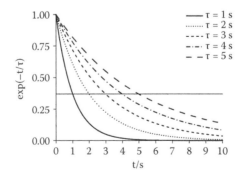

FIGURE 2.22
Exponential decay with a time constant τ, so that $b = 1/\tau$ in Equation 2.13. Five different values for τ are shown. In all cases the exponential function has a value of 0.37 when $t = \tau$; this value is shown by the fine horizontal line.

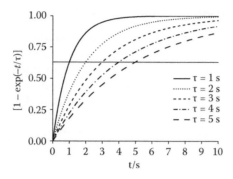

FIGURE 2.23
The function [1 − exp(−*t*/τ)] shown plotted for five different values of the time constant τ. For all values of τ the function has a value of 0.63 when *t* = τ; this value is shown by the fine horizontal line.

affected by the value of the time constant τ. Rapid decay is associated with smaller values of the time constant.

A second function that is commonly plotted is [1 − exp(−*t*/τ)]. This is a function that grows from a value of zero at *t* = 0 to a value of 1 for large *t* (Figure 2.23). For this function, at time *t* equal to the time constant, the signal has risen to 0.63 of its final value.

Question 2.5

y = exp(−*t*/τ), where τ is the time constant.

    (a) What is the quantitative definition of τ?
    (b) When *t* = τ, what proportion of the original signal remains?
    (c) When *t* = 2τ, what proportion of the original signal remains?
    (d) When *t* = 5τ, what proportion of the original signal remains?

Answer

τ is the time taken to reach a point on the curve with the value e^{-1}. When *t* = τ, 0.37 of the original signal remains. The corresponding proportion for *t* = 2τ is 0.135, and 6.7×10^{-3} for *t* = 5τ. It is sometimes taken as a rule of thumb that the signal can be considered to be negligible after a time equivalent to five time constants has elapsed, and this approximation may be justified using the calculation in part (d).

Question 2.6

y = exp(−*t*/τ), where τ is the time constant. Will a curve with a large time constant lie above or below one with a smaller time constant?

Answer

The curve with the larger time constant will lie above the curve with the smaller time constant. See, for example, the curves plotted in Figure 2.22.

Question 2.7

Consider the function $y = [1 - \exp(-t/\tau)]$.

  (a) What is the value of y when $t = \tau$?
  (b) If $t = 5\tau$, is the value of y greater or less than when $t = \tau$?

Answer

When $t = \tau$, $y = 0.63$. The value of y is greater at $t = 5\tau$ than at $t = \tau$; this is expected, as the expression describes a function that increases from 0 to 1.

Question 2.8

$A = 1 - \exp(-t/\tau_1)$, $B = 1 - \exp(-t/\tau_2)$, and $\tau_1 < \tau_2$. On the same axes, sketch the curves A and B.

Answer

Your sketch should have a horizontal t axis and show two curves, shaped as in Figure 2.23, which both increase in value from 0 to 1. Curve A, with the shorter time constant, will have a larger value than curve B at all values of t, and will reach the maximum value 1 at a smaller value for t than does the B curve.

2.2.8.2 Exponential Envelope

If a periodic function is multiplied by an exponential decay, the periodic function maintains its frequency, but its amplitude is modulated by an envelope shaped like the decay curve and its mirror image (Figure 2.24).

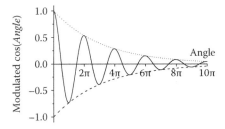

FIGURE 2.24
The result (solid curve) of multiplying a periodic signal by an exponential decay function (dotted). The amplitude of the periodic function is modulated by an envelope defined by the shape of the decay curve.

This is relevant where the two processes are happening at the same time; the overall result is the product of the two functions.

2.2.9 Links to Later Chapters

Table 2.6 indicates where the mathematics outlined in this chapter is used later in the book.

TABLE 2.6

The Chapters in Which Mathematics Outlined in This Chapter May Be Useful

Mathematical Background Section	Applied Elsewhere in Book
2.2.1.2–2.2.1.3	The trigonometric relationships are used in Chapters 3 and 10 (gradient echo).
2.2.1.4	Degrees and radians are both used throughout the book.
2.2.1.5	Phase differences are important in Chapters 7 (phase encoding) and 10 (flow). See also Section 2.2.5.
2.2.2.2	The principle of splitting a vector into its components is an important concept for MRI, and it is an operation essential for understanding Chapters 3, 4, and 8.
2.2.2.3	Vector addition is important for understanding the basic concept underlying the classical description of MRI presented in Chapter 3.
2.2.3	Three-dimensional vectors are used repeatedly in MRI. You may need to refer to this section when reading Chapters 3 and 4.
2.2.4	The relationship between linear and angular frequency is an important one that is necessary when applying the equations that are associated with both the main and gradient magnetic fields in MRI. You may need to refer to this section when working on Chapters 3 through 7.
2.2.5	The phase diagram is used extensively in MRI to illustrate phase differences. It is particularly relevant to the understanding of phase encoding (Chapter 7) and the principles of the spin-echo sequence (Chapter 8).
2.2.6	Sampling and aliasing are relevant to MRI because they can explain the occurrence of some imaging artifacts.
2.2.7	Fourier analysis is used to convert measured MRI signals into images (Chapters 6 and 7). The role of high and low frequencies in a signal is relevant to the data matrix (Chapter 7).
2.2.8	The exponential function underlies the relaxation mechanisms in MRI (Chapter 4). Consequently, it is important for pulse sequences (saturation recovery in Chapter 4 and inversion recovery in Chapter 10) and imaging pulse sequences (Chapters 8 and 10).
2.2.8.2	Modulation of a signal by an exponential function is associated with free induction decay (Chapter 3).

2.3 Physics for MRI

The physics outlined in this section is most relevant to the understanding of Chapter 3.

2.3.1 Atomic Structure

The atom is made up of a *nucleus* surrounded by orbiting *electrons*. Electrons are small particles carrying a negative charge. The nucleus contains two types of particle: *protons* and *neutrons*. Protons have a positive charge and neutrons have no charge. The unit of electrical charge is the coulomb (C), which corresponds with the charge on 6.24×10^{18} electrons. The charge on a single electron is -1.6×10^{-19} C, and the charge on a single proton is $+1.6 \times 10^{-19}$ C. The atom as a whole has the same number of electrons as protons and so has no charge.

Each element has a characteristic number of protons in the nucleus. The number of protons in the nucleus is the *atomic number* of an atom, which has the symbol Z. The atomic number is fixed for a particular element; for example, carbon always has $Z = 6$, hydrogen has $Z = 1$.

Question 2.9

Which of the following reside in the nucleus of an atom? Electron, neutron, proton, boron, photon?

Answer

The neutron and proton can be found in the nucleus of an atom.

2.3.2 Electricity

Any stationary electric charge is surrounded by a continuous *electric field*. Electric field lines radiate out from positive charges and into negative charges. The strength of the electric field is represented by the symbol E and measured in volts per meter.

- The strength of an electric field drops off with distance, r, from the charge according to $1/r^2$, i.e., an inverse square law.
- The strength of an electric field is directly proportional to the amount of charge.
- Like charges repel, and unlike charges attract.

The *electric potential* is a measure of the energy required to move an electron in an electric field. Potential has the symbol V and units of volts

(V). An *electric current* is a flow of electric charge. The unit of current is the ampere (A), which represents a flow of charge of 1 coulomb per second. Note that when a current flows in an electric circuit, it usually encounters components that impede its flow and result in energy being converted to other forms:

- Resistance leads to generation of heat.
- Inductance leads to generation of magnetic fields.
- Capacitance leads to generation of electric fields.

2.3.3 Magnetism

Any moving charge generates a magnetic field. Matter is made up of atoms, which incorporate moving charges (electrons), so all matter exhibits some degree of magnetism and will respond to a magnetic field. The degree of magnetism exhibited by a large-scale sample of material depends on the degree of organization of the motion of the electrons. In permanent magnets there is a high degree of organization.

Magnetic susceptibility is the extent to which a material becomes magnetized when placed in a magnetic field. The symbol for magnetic susceptibility is χ, and magnetic susceptibility is a number with no units.

- Materials with positive susceptibility have a field induced within them that results in a total field that is greater than the strength of the magnetic field in which they are placed (Figure 2.25b). Such materials are described as paramagnetic.

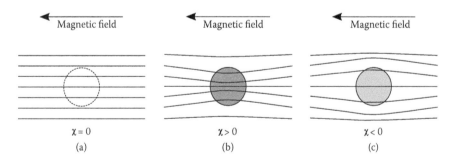

FIGURE 2.25
Magnetic susceptibility. (a) The magnetic susceptibility of a vacuum is zero, so magnetic field lines are unaffected. (b) Materials with positive susceptibility are called paramagnetic. The field induced within the material is in the same direction as the external field, and the total field within the material is greater than the external field, indicated by the denser field lines. (c) Materials with negative susceptibility are called diamagnetic. The field induced within the material is in the opposite direction to the external field, so the field within the material is smaller than the external field, indicated by the less dense field lines.

TABLE 2.7

The Three Types of Magnetic Material

Type of Material	Effect	Susceptibility	Examples
Diamagnetic	Induced magnetic field opposes external field	$\chi < 0$	Bismuth, copper
Paramagnetic	Induced magnetic field enhances external field	$\chi > 0$	Aluminum, manganese, oxygen, platinum, tungsten
Ferromagnetic	Induced magnetic field strongly enhances external field. Very strongly paramagnetic materials are called *superparamagnetic*	$\chi > 1$	Easily magnetized materials such as cobalt, iron, and nickel. Also gadolinium, which is used in MR contrast agents

- Materials with negative susceptibility have an induced field that opposes the field in which the material has been placed, so the total field within the material is lower than the field in which they are placed (Figure 2.25c). Such materials are described as diamagnetic.

Magnetic materials can be classified into three types depending on their magnetic susceptibility, as shown in Table 2.7.

We have previously seen that electric charge comes in small, discrete units, as carried by the electron or proton. The charge can be either positive or negative. In contrast, magnetism always comes in pairs, never singly. Both a north pole and a south pole are always present, and such a pair is called a *magnetic dipole*. Magnetic field lines run out from the north pole and into the south pole; like poles repel each other and unlike poles attract.

Just as is the case for electric fields, the strength of a magnetic field falls off with distance according to an inverse square law. The symbol for *magnetic field strength* is B, and the unit of magnetic field strength is the tesla (T). Note that sometimes the unit gauss (G) is used instead. It is not the correct SI unit, but can be more convenient for describing smaller fields, because 1 T = 10,000 G.

Question 2.10

Which of the following statements are true for diamagnetic materials?

(a) The magnetic susceptibility is less than 0.
(b) The magnetic susceptibility is greater than 1.
(c) Examples include aluminum, manganese, oxygen, platinum, and tungsten.
(d) Examples include bismuth and copper.
(e) The field within the material is lower than the main field.

Answer

The true statements are (a), (d), and (e).

2.3.4 Electromagnetism

We have defined electric and magnetic fields separately, but the two fields usually occur together and represent two different ways to describe an *electromagnetic* field. For example, a stationary electric charge possesses an electric field, but whenever an electric charge moves there is also a magnetic field in addition to the electric field. This means that there will be a magnetic field associated with the flow of electrical charge (a current), for example, in a wire or a coil. This is the principle that supports the use of coils for generating a magnetic field in MRI.

In the same way that a magnetic field is always associated with a moving charge, an electric current is always present when there is a moving magnetic field, or if a material moves through a static magnetic field. *Eddy currents* are induced in this way in the iron cores of a coil, rather than in the coil windings. Such eddy currents result in the production of heat.

2.3.5 Electromagnetic Radiation

Electromagnetic radiation arises from an oscillating electric field (E) or from an oscillating magnetic field (B). Whichever one is present, the other is always present too and is induced perpendicular to the first field. The velocity of propagation of all electromagnetic radiation is the velocity of light (c), and the direction of this propagation is one that is orthogonal to both fields (Figure 2.26). Different types of electromagnetic radiation are characterized by their frequency of oscillation (Figure 2.27). For example, x-rays and radio waves are both types of electromagnetic radiation, but they have different frequencies. Radio frequency (RF) radiation is a vital part of MRI.

IONIZATION

Ionization occurs in other medical imaging modalities, such as x-ray imaging and radionuclide imaging, but is not a hazard associated with MRI. The atom is made up of a nucleus surrounded by orbiting electrons, and the orbits fill up with electrons from the innermost orbit outward. The innermost orbit contains two electrons; there are eight in the next one out, and there is a preferred number for the other orbits. The outermost orbit never has more than eight electrons. If, for any reason, an electron is lost from an inner orbit,

---

one will drop down from further out, to take its place and ensure maximum occupancy of the inner orbits. An atom is said to be *ionized* if one of its orbiting electrons is removed. If this happens, the atom is no longer neutral and it is called a positive ion. Electrons can be removed in this way if energy is supplied to the atom. The energy of x-rays and gamma radiation is sufficient to remove an electron from the innermost orbit of an atom, and bring about ionization. However, the energy of radio frequency radiation as used in MRI is not sufficient to cause ionization. Ionization causes tissue damage by inducing chemical changes in tissue. Safety concerns with MRI are associated instead with the mechanical and biological effects of the three magnetic fields used in MRI: the large static field, time-varying gradient fields, and the RF fields.

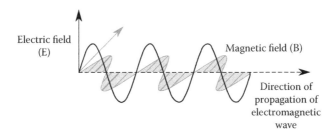

FIGURE 2.26
The electromagnetic wave is made up of both an electric field and a magnetic field. The two fields oscillate in directions perpendicular to each other, which are also both perpendicular to the direction of propagation. In this diagram the electric (black) and magnetic (gray) fields are shown in phase with each other.

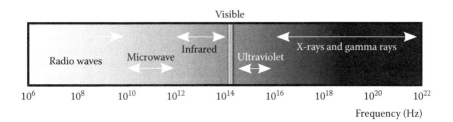

FIGURE 2.27
The electromagnetic spectrum. The frequency ranges associated with some of the main types of electromagnetic radiation are indicated. In some illustrations of the electromagnetic spectrum, scales are shown not only for frequency but also for wavelength and energy. This is possible because there is a fixed relationship between the frequency of the radiation and its wavelength or energy. High energy corresponds with high frequencies and short wavelengths.

Question 2.11

Which of the following statements are true concerning electromagnetic radiation?

- (a) The frequency of x-rays is higher than the frequency of RF radiation.
- (b) Visible light has a lower frequency than RF radiation.
- (c) The velocity of electromagnetic radiation in a vacuum is a constant.
- (d) The electric and magnetic fields are perpendicular to one another.
- (e) The electric and magnetic fields are parallel to one another.

Answer

The true statements are (a), (c), and (d).

2.4 Imaging for MRI

Magnetic resonance imaging, as the name suggests, is a technique for producing images. Several concepts associated with digital images are assumed in the later chapters of this book, so these ideas are introduced here.

2.4.1 Digital Image, Pixel, and Voxel

A digital image is an array of numbers. This is generally a rectangular array, and often square: a common matrix size in MRI is 256 × 256. Each number represents the brightness at a location in the subject, and when the numbers are displayed in the correct order, we see an image. Each of the numbers is known as a picture element, or *pixel*. The image size, sometimes called the *matrix*, is expressed in terms of the number of pixels in the square or rectangular array, for example, 256 × 192 pixels. A *voxel* (volume element) is the three-dimensional analog of a pixel, and has a value associated with a small volume of tissue. *Pixel size* is usually expressed in millimeters and indicates the size of the pixel within the imaged subject.

2.4.2 Gray-Scale Resolution

The gray-scale, or brightness, resolution in an image is determined by how many brightness values can be stored. In MRI it is common to assign the pixel value to one of 4,096 discrete values in the range from 0 to 4,095. An image with 4,096 possible gray values is called a 12-bit image, because data are stored in a binary format, and 2^{12} is equal to 4,096 (see "Bit Depth" box).

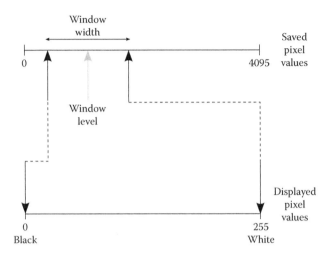

FIGURE 2.28
Window width and level. In this example, the saved pixel values range from 0 to 4,095. Using a monitor that displays values from 0 to 255, a range of gray values (window width) centered around a selected value (window level) is chosen. These values are then scaled into the range from 0 to 255. If the range of saved pixel values is the same as the display range, the window width can be used to select a smaller range of gray values for display. This will enhance the contrast seen in the displayed image.

Most image displays, however, can show only 256 discrete values (8 bits), running from 0 to 255, so it is usual to *window* the data for display.

2.4.2.1 Windowing

Windowing involves the selection of a range of pixel values that are to be displayed using the available number of gray values. The window is defined in terms of its width and level, as illustrated in Figure 2.28. The window does not affect the stored data, which retain their full gray-scale resolution.

BIT DEPTH

The fundamental unit in computing is the bit, which has two possible states: on or off. With a group of 8 bits, there are 2^8 possible different arrangements of on- and off-bits. As a result, 256 (2^8) different numbers can be stored or displayed using 8 bits. Similarly, 2^{12} (4,096) possible values can be stored using 12 bits, and 2^{16} (65,536) with 16 bits.

2.4.3 Spatial Resolution

The spatial resolution indicates the extent to which closely spaced objects can be distinguished from each other in the acquired image. The spatial resolution depends partly on the performance of the imaging system, and includes, for example, the effect of noise, but it is also strongly related to the matrix size and field of view. The relationship between the field of view and matrix size (Figure 2.29) means that the highest spatial resolution is achieved by having a small field of view together with a large matrix size, so that each pixel covers only a small area in the subject.

Note that once the image has been acquired, increasing the matrix size of the image and interpolating new pixel values does not increase the spatial resolution, although it may improve the appearance of the image. Interpolation works on data that have already been acquired, and so the acquisition resolution remains unchanged.

2.4.4 Image Contrast

It is possible to distinguish different structures in an image because of differences in image brightness that mean that they can be seen against their surroundings. This degree of visibility is called the *image contrast*. In an image with low contrast it is difficult to distinguish different features, even when they are large enough to be spatially resolved. It is a feature of MRI that it is possible to adjust acquisition sequences in order to maximize the difference in brightness between selected tissue types.

When viewing an image it is common to perform interactive contrast enhancement on the display. The simplest way to go about this is to use windowing (Section 2.4.2.1) to spread a small range of gray values across the full display range.

2.4.5 Signal-to-Noise Ratio

Noise in an image is a random variation in gray values, which gives it a speckled appearance. Both spatial and gray-scale resolution are affected by the presence of noise. The *signal-to-noise ratio* (SNR) is a measure of the noise present in an image. SNR is given by the mean signal in the area of interest divided by the measured standard deviation of the noisy image background. SNR is high in good-quality images where features can be easily distinguished from the surroundings (Figure 2.30).

The *contrast-to-noise ratio* (CNR) can be a more useful measure than SNR, because it considers how tissue differentiation has been affected by noise. CNR is given by the difference in the mean signal in two areas of interest divided by the measured standard deviation of the noisy image background.

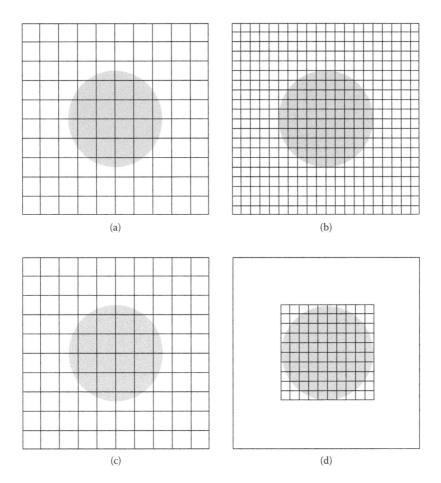

FIGURE 2.29
The relationships among field of view (FOV), matrix size, and spatial resolution. (a) and (b) have the same FOV (indicated by the square boundary), but the larger matrix in (b) leads to smaller pixels than for (a), and thus higher spatial resolution. (c) and (d) have the same matrix size, but the smaller FOV in (d) leads to smaller pixels than for (c), and thus higher spatial resolution.

2.4.6 Geometrical Image Properties

Image nonuniformity, or inhomogeneity, can be either spatial or gray scale in character. Spatial distortion affects the geometry of the image so that, for example, a straight line in the object will appear bent or at a different angle in the image. Gray-scale effects arise when different parts of an object with uniform properties have different gray values in the image.

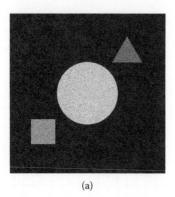 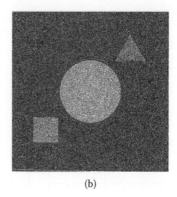

(a)   (b)

FIGURE 2.30

Illustration of the signal-to-noise ratio (SNR) for an image affected by different amounts of noise. (a) This image has less noise than the other image. The SNR for the large circle is about 14. (b) In this noisier image the SNR for the large circle is about 3. Note that the mean signal in the large circle is lower here than in (a), and the background signal has a larger range of values; both contribute to the reduction in SNR.

Question 2.12

Which of the following statements are true?

(a) Pixel is a word derived from a picture element.
(b) Vixel is a three-dimensional pixel.
(c) If the pixel size in an image is 0.5 mm, then the image has lower spatial resolution than an image where the pixel size is 1.5 mm.
(d) The smaller the pixel size, then the higher the gray-scale resolution.
(e) If a larger matrix size is used within the same field of view, then the spatial resolution is improved.

Answer

The correct statements are (a) and (e).

2.5 Clinical Imaging Terms for MRI

Definitions of clinical imaging terms that are used later in the book are given in Table 2.8.

TABLE 2.8

Clinical Imaging Terms

Term	Definition
Angiogram	An image that shows blood vessels.
Anterior–posterior direction	Direction running from front to back of the body.
Axial image	An image in a plane that divides the superior and inferior parts of the body (Figure 2.31a), perpendicular to the long axis of the body. Also known as a transverse image.
Coronal image	An image in a plane that divides the anterior and posterior parts of the body (Figure 2.31c).
Cranio-caudal direction	Direction running from head to toe.
Isocenter	The zero point of the three-dimensional coordinate system.
Prone	Lying on the front with face downward.
Sagittal image	An image in a plane that divides the left and right sides of the body (Figure 2.31b).
Supine	Lying on the back with face upward.
Transverse image	An image in a plane that divides the superior and inferior parts of the body (Figure 2.31a).

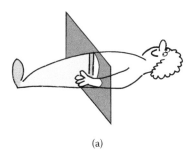

(a)

FIGURE 2.31
Illustration of the three orthogonal anatomical planes. (a) The transverse plane, which divides the body into upper (head) and lower parts. (b) The sagittal plane, which divides the body from side to side. (c) The coronal plane, which divides the anterior and posterior of the body.

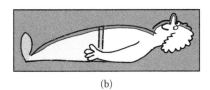

(b)

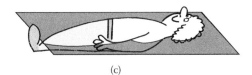

(c)

2.6 Chapter Summary

Background information about mathematics, physics, digital imaging, and clinical imaging terms was covered in this chapter. The chapter may be used for reference when the terms are encountered in later chapters. Those unfamiliar or out of practice with the material can also use the chapter, together with the first 50 multiple-choice questions in Chapter 11, for revision purposes.

3

Basic Physical Principles

In this chapter the physical processes that lead to the generation of a signal in magnetic resonance imaging (MRI) are introduced. Although a complete description can be very complex, many of the details are not necessary for developing an understanding of how the processes are used in MRI. The description given here therefore is limited to the essential core of information. All MRI systems have three basic components, and these are introduced in this chapter. It is shown how the need for each of the three components can be linked to the physical processes that are involved. The first of the interactive programs—the *Rotating Frame of Reference* program—is used in this chapter.

3.1 Learning Outcomes

When you have worked through this chapter, you should:

- Be able to explain the origin of the signal that is acquired in MRI using descriptions from classical physics.
- Be able to list the three main components of an MRI system and understand how the roles of the components relate to the basic physical principles of MRI.

3.2 Spins and the Net Magnetization Vector

The nuclei of some atoms possess a small magnetic field, which arises because the nucleus has a charge and is spinning (Figure 3.1). Nuclei with a small magnetic field are often described as *spins*. Millions of nuclei are present in a sample of material, and the magnetic field resulting from the combined effect of all the small magnetic fields associated with the spinning, charged nuclei can be described in terms of a vector, the *net magnetization vector*. The net magnetization is zero when there is no external

FIGURE 3.1
The nucleus has a spinning electrical charge, which acts like an electrical current loop and leads to the presence of a magnetic field.

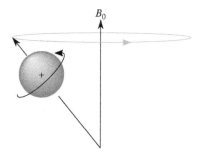

FIGURE 3.2
Precessional motion about the direction of an external magnetic field, B_0.

magnetic field present, because the average effect of a large number of randomly oriented spins is zero.

When such charged, spinning nuclei are placed in a strong uniform magnetic field (B_0), the motion of each nucleus changes. Each nucleus was already spinning about its own axis, so the effect of the external field is a complex motion about the direction of the external field (Figure 3.2). This motion is called *precession* (see "Precession" box). The frequency of precession depends on the characteristics of the nucleus and the strength of the magnetic field, and we shall consider this relationship in the next section. The precessional motion can take up one of two preferred states. The two states can be visualized by considering the small magnetic fields associated with spins as follows:

- The small magnetic fields are aligned parallel to the direction of the external magnetic field (B_0).
- The small magnetic fields are aligned antiparallel to the direction of the external magnetic field (B_0).

These two states may be called *spin-up* and *spin-down* (Figure 3.3). Although this visualization is not a completely accurate description,

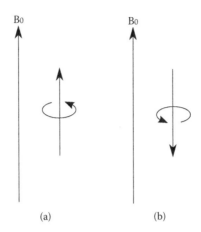

FIGURE 3.3
The simplified spin-up, spin-down notation. (a) In the spin-up state, precession is about a direction parallel to the external field B_0. (b) In the spin-down state, precession is about a direction antiparallel to the external field B_0.

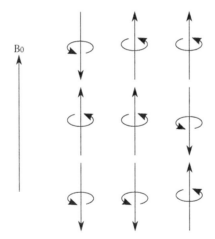

FIGURE 3.4
Shortly after the external magnetic field B_0 is applied, there are more spins in the spin-up state than there are in the spin-down state. As a result, the net magnetization vector, which is the vector sum of the many small magnetic fields, has a positive value in the direction of B_0.

it works well when developing an understanding of the principles (see "Spin-Up and Spin-Down States" box). Shortly after an external magnetic field is applied, there are always more spins in the spin-up state than in the spin-down state (Figure 3.4). As a result, the net magnetization vector is not zero, as it was when there was no external field present.

PRECESSION

Precession is the motion made by a spinning top or gyroscope in a gravitational field. The top spins around its own axis, and if it is pushed over from its vertical alignment, its upper end traces out a circular path around the vertical direction.

One example of a nucleus that has a spin that exhibits this behavior in a magnetic field is the nucleus of the hydrogen atom, which consists of a single proton. The hydrogen atom is abundant in the human body because the human body is largely composed of water, and because the atom is present in fat. It is because of this abundance in the body that the hydrogen nucleus is used in clinical magnetic resonance imaging. Other nuclei that behave in the same way in a magnetic field are less abundant, and this means that the detection sensitivity is much lower, for example, the detection sensitivity for the hydrogen nucleus is about 1,000 times greater than for sodium (23 Na) and 700 times greater than for phosphorous (31 P) [1].

SPIN-UP AND SPIN-DOWN STATES

A full explanation of why two, and only two, states should exist requires quantum mechanics. The two states are both stable, so the reason that the spin-up state is preferred to the spin-down state is related to the energy required. Slightly more energy is required for the spin-down state. The effect of adding energy is to move protons from the lower-energy spin-up state to the higher-energy spin-down state.

3.3 The Larmor Equation

The existence of a net magnetization vector when a strong magnetic field is present can lead to an emitted signal, which can be measured and eventually used to generate an image. To understand how such a signal arises, it is necessary to consider the frequency of precession associated with the magnetic field.

3.3.1 Larmor Frequency

The Larmor frequency is the rate of precession of a spin in an external magnetic field. The frequency is defined in the *Larmor equation*

$$\omega_0 = \gamma B_0 \tag{3.1}$$

where ω_0 is the angular precessional frequency, measured in radians per second (rad s^{-1}), B_0 is the external magnetic field in tesla (T), and γ is the *gyromagnetic ratio*, measured in rad s^{-1} T^{-1}.

Linear frequency, f, measured in hertz, is more often used than angular frequency. The relation between angular and linear frequency is

$$\omega = 2\pi f \tag{3.2}$$

This leads to an alternative way of writing the Larmor equation:

$$f_0 = \frac{\gamma}{2\pi} B_0 \tag{3.3}$$

where f_0 is the linear precessional frequency, measured in hertz, B_0 is the external magnetic field in tesla (T), and $\gamma/2\pi$ is the gyromagnetic ratio. The gyromagnetic ratio in this case is measured using units of linear frequency, for example, in Hz T^{-1}, instead of units of angular frequency. For MRI the high frequencies involved mean that the gyromagnetic ratio is usually given in the units MHz T^{-1}.

To distinguish between the two different ways of expressing the gyromagnetic ratio, it is best to write the linear frequency alternative, $\gamma/2\pi$, using a symbol that indicates the presence of the 2π. One option is $\bar{\gamma}$, pronounced gamma bar, where the bar indicates the division by 2π. Equation 3.3 may then be rewritten as follows:

$$f_0 = \bar{\gamma} B_0 \tag{3.4}$$

The symbol $\bar{\gamma}$ is used in this book and others, but its use is not widespread. When the gyromagnetic ratio is expressed in MHz T^{-1} (and not in rad s^{-1} T^{-1}), it can be assumed that the value given represents $\gamma/2\pi$, and the version of the Larmor equation using linear frequency (Equation 3.3) is the appropriate one to use.

3.3.2 Gyromagnetic Ratio

The gyromagnetic ratio is a constant for a particular atomic nucleus. For the hydrogen nucleus,

$$\gamma = 42.6 \text{ MHz T}^{-1} \ (\gamma = 2.67 \times 10^8 \text{ rad s}^{-1} \text{ T}^{-1})$$

The differing gyromagnetic ratios for nuclei of different atoms mean that each different nucleus precesses at its own characteristic frequency in a strong magnetic field. Working with a particular frequency means that it is possible to concentrate on the behavior of the hydrogen nucleus and ignore the contributions from the other nuclei that also precess in a strong magnetic field.

Worked Example

Calculate the precessional frequency for the hydrogen nucleus for magnetic field strengths of 0.5 T, 1.0 T, and 3.0 T. The gyromagnetic ratio is 42.56 MHz T^{-1}.

This question is addressed using the Larmor equation, which links precessional frequency to magnetic field strength. The gyromagnetic ratio is given in the question in units of linear frequency, so the version of the Larmor equation to use is Equation 3.4: $f_0 = \gamma B_0$.

By substituting in this expression, separately, for the three different magnetic field strengths, it can be calculated that the precessional frequency for the hydrogen nucleus at 0.5 T is 21.28 MHz; at 1.0 T, 42.56 MHz; and at 3.0 T, 127.68 MHz.

Note how the linear relationship between frequency and magnetic field strength (Equation 3.4) means that it is relatively simple to deduce the precessional frequency given a change in the magnetic field strength. For example, if the precessional frequency is known at a given field strength, and that field strength is doubled, then the precessional frequency is doubled too.

Question 3.1

Calculate the precessional frequency for the hydrogen nucleus in a magnetic field strength of 1.5 T. The gyromagnetic ratio is 42.56 MHz T^{-1}.

Answer

Question 3.1 is very similar to the worked example, and by substituting in Equation 3.4, it is found that the precessional frequency for the hydrogen nucleus in a magnetic field strength of 1.5 T is 63.84 MHz.

Question 3.2

Calculate the precessional frequency, as both an angular frequency and a linear frequency, for the hydrogen nucleus in a magnetic field strength of 2.0 T. The gyromagnetic ratio is 2.67×10^8 rad s^{-1} T^{-1}.

Answer

In Question 3.2 the gyromagnetic ratio is given in units of rad s^{-1} T^{-1}, and this suggests that the version of the Larmor equation to use is Equation 3.1: $\omega_0 = \gamma B_0$. Substitution gives the angular precessional frequency at 2.0 T to be 5.34×10^8 rad s^{-1}. The conversion from angular to linear frequency is accomplished using Equation 3.2, and the answer in linear frequency is found to be approximately 85.0 MHz (to three significant figures). The answer can be cross-checked by using Equation 3.4 and assuming, in that case, that the gyromagnetic ratio is 42.56 MHz T^{-1}.

THE MAIN MAGNETIC FIELD IN AN MR SYSTEM

Every MRI system has a magnet that has a strong magnetic field. Such a field is necessary so that the hydrogen nuclei placed in it precess about the direction of the field at the Larmor frequency. Field strengths in MR systems are classified as being low/ultralow (from 0.02 T to 0.2 T), medium (from 0.2 T to 1.0 T), or high (1.0 T and above). For comparison, the electromagnets used for working with scrap metal have magnetic fields of up to 2.0 T, a refrigerator has a magnetic field of about 10 mT, and the magnetic field of the Earth at the equator is 30 μT.

The type of magnet used to generate the field is different for the three ranges. Resistive magnets are used for low and medium fields, permanent magnets provide medium fields, and superconducting (cryogenic) magnets are used for high fields. High magnetic field strengths give a stronger signal and lead to higher signal-to-noise ratios in the images, but this advantage is partly offset by the presence of imaging artifacts and the higher capital cost compared with lower field strengths.

The magnetic field within the imaging volume must be homogeneous. Magnetic field variations are expressed in parts per million (ppm) across a stated field of view. A superconducting magnet may have a variation of 15 ppm.

3.4 Nuclear Magnetic Resonance

In the discussion so far we have reached a point where hydrogen nuclei are precessing at a known frequency (the Larmor frequency), which depends on the strength of the applied magnetic field. The next step is to consider the absorption of energy by the system. Energy will only be absorbed by the system if it has the same frequency as the precessing nuclei, that is, the Larmor frequency. Energy at any other frequency will not be absorbed. This absorption of energy at the resonant frequency is called *nuclear magnetic resonance.* The Larmor frequency of the hydrogen nucleus, at the field strengths used in MRI, corresponds with the radio frequency (RF) range in the electromagnetic spectrum. The absorption of the RF radiation has an effect on the net magnetization vector and causes it to move toward the xy plane.

The need to supply energy at radio frequencies explains why the second main component of an MR system comprises coils that can transmit and receive RF pulses.

3.4.1 Visualization of the Effect of Radio Frequency (RF) Pulses

The coordinate system used in MRI is shown in Figure 3.5. The external magnetic field is always aligned with the z direction and the net magnetization vector, M, precesses about this direction. The net magnetization vector can be resolved into two components, as shown in Figure 3.6. M_z is the component in the direction of the external field B_0.

The nuclear magnetic resonance principle means that if energy is supplied at the Larmor frequency of a particular nucleus, then that energy will be absorbed. We have seen that without the additional energy, there are more spins in the spin-up state. When energy is supplied, the effect is

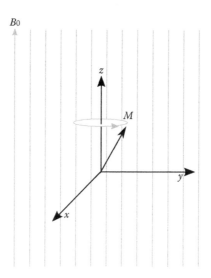

FIGURE 3.5

The coordinate system used when discussing MRI. The main magnetic field B_0 is applied in the z direction, and the signal is measured in the xy plane. The net magnetization vector M arises from the combined effect of the small magnetic fields possessed by millions of nuclei. The vector may be resolved into two components: one in the z direction and one in the xy plane.

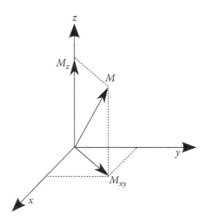

FIGURE 3.6
The longitudinal (M_z) and transverse (M_{xy}) components of magnetization.

to move some of the spins that are in the spin-up state into the spin-down state. As more energy is supplied, the situation changes from there being an excess of spins in the positive z direction (Figure 3.7a) to having an excess in the negative z direction (Figure 3.7c). In between, there will be a condition when equal numbers of spins are in the two different states, and there is no net magnetization in the z direction (Figure 3.7b).

It is convenient to visualize the change in occupancy of the two states in terms of the tipping of the net magnetization vector, and this is a method of describing the process that is used extensively in MRI. The net magnetization vector starts off aligned with the positive z direction, and the component in the z direction is large and equal to the vector magnitude (Figure 3.8a). As energy is supplied, it tips toward the xy plane, so that although the magnitude of the vector has not changed, the component in the z direction has decreased (Figure 3.8b). When the vector has tipped through 90°, the z component is zero; this corresponds with an equal number of spins being in the spin-up and spin-down states (Figure 3.8c). Further energy supply (still at the correct frequency) increases the angle further, and the z component of the magnetization vector grows in the negative z direction, corresponding with the higher number of spins now in the spin-down state (Figure 3.8d). When the vector has tipped through 180°, the z component is at its maximum negative value (Figure 3.8e).

3.5 Longitudinal and Transverse Magnetization

So far, the emphasis has been on the z component of the net magnetization vector, and its relationship with the occupancy of the two favored states.

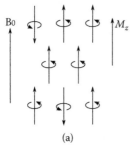

(a)

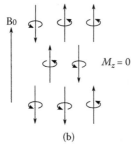

(b)

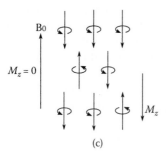

(c)

FIGURE 3.7
As energy is added at the Larmor frequency, spins move from the spin-up to the spin-down state. (a) More spins are in the spin-up state, and the z component of the net magnetization vector (M_z) is in the positive z direction. (b) The number of spins in the spin-up and spin-down states are equal. M_z is zero. (c) More spins are in the spin-down state, and the z component of the net magnetization vector (M_z) is in the negative z direction.

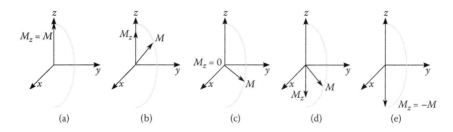

(a)   (b)   (c)   (d)   (e)

FIGURE 3.8
Tipping of the net magnetization vector M. The gray dashed line indicates the path of the tip of the vector, ignoring the effect of precessional motion. As energy is supplied at the Larmor frequency, the changes in the size of the z-component M_z can be seen. The vector M tips from (a) alignment with the positive z axis to (e) alignment with the negative z axis via intermediate positions shown in (b), (c), and (d). Figure 3.7a,b,c correspond with (a), (c), and (e) in this figure.

The *z* component is given the symbol M_z and is known as the *longitudinal magnetization*. For MRI the component at right angles to this is also of great importance, because the MR signal is measured in the plane at right angles to the applied external field. The component at right angles to the *z* component is the *xy* component of magnetization, M_{xy}, and it is called the *transverse magnetization* (Figure 3.6).

3.5.1 Full Magnetization and Saturation

When the M_z component of magnetization is at its maximum (*M* is aligned with the *z* axis) the system is said to be *fully magnetized*. *Saturation* is the opposite state to full magnetization. A saturated system exists when the M_z component of magnetization is zero, which is the case after a 90° flip for which the *flip angle* between *M* and the *z* axis is 90° (Figure 3.9).

3.5.2 Free Induction Decay Signal

Figure 3.9 is simplified, because the precessional motion of *M* is not shown. The precession of *M* means that, for a given angle of precession, the tip of the M_{xy} component describes a circle in the *xy* plane (Figure 3.10). For measurement of the signal, the *xy* component is further resolved into separate *x* and *y* components (sometimes called the real and imaginary, or in-phase and in-quadrature, components). Each component varies as a sine function because of the rotation of M_{xy}. The signal is also reduced by decay processes that we shall consider in Chapter 4, and these processes mean that the sinusoidal signal is modulated by an exponential function (Figure 3.11). The measured signal is known as *free induction decay* (FID). The FID signal decays very quickly, and we shall see in Chapter 8 that the signal measured in practice is not a direct measurement of the FID signal.

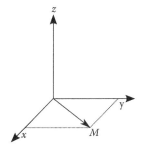

FIGURE 3.9
The system is said to be saturated when the net magnetization vector lies in the *xy* plane so that the M_z component is zero.

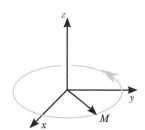

FIGURE 3.10
The precessional motion means that the tip of the *xy* component of the net magnetization vector describes a circular path in the *xy* plane.

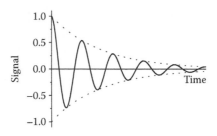

FIGURE 3.11

The solid line shows the free induction decay (FID) signal, which is measured in the xy plane. The dotted lines indicate how the signal size is modulated by the exponential decay function $\exp(-t/T_2^*)$ (Section 4.2.4).

THE RF COILS IN AN MR SYSTEM

There are two functions for RF coils in an MR system: transmitting and receiving. Transmission RF coils are used to generate the pulses that tip the net magnetization vector from its equilibrium position. Receiver RF coils are used to acquire the emitted signal. All RF coils are composed of a loop, or multiple loops, of wire, and can be further classified as being either body or surface coils. Body coils are usually built into the bore of the main field magnet, and the same coil is capable of both transmit and receive operations. Surface coils are designed as receiver coils, and to give higher signal-to-noise ratio images for a particular anatomical site. Phased array coils are used to cover a large field of view. The array is made up of several surface coils, each with its own receiver.

The field generated or measured by an RF coil is always perpendicular to the main magnetic field of the system. The transmitted magnetic field associated with the RF pulse is sometimes called B_1 to distinguish it from the main B_0 magnetic field.

3.6 Rotating Frame of Reference

Visualization of the tipping net magnetization vector is very helpful for understanding MRI, but the visualization is hampered by the precession of the net magnetization vector. The path drawn by the tip of the vector as energy is supplied is a spiral path, because the vector is moving toward the xy plane and also precessing. Do not worry if you find this spiraling motion hard to visualize; the exercises that follow using the Rotating Frame of Reference program will make things clear. The *rotating frame of*

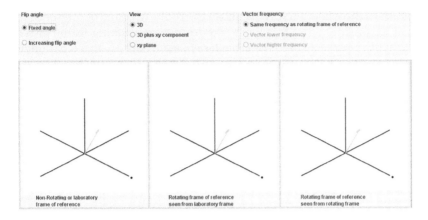

FIGURE 3.12
The appearance of the rotating frame of reference program when it is first run. The parameter-setting area is at the top, and three frames showing different views of the laboratory and rotating frames of reference appear underneath.

reference is a way of removing the confusion of the precessional motion. At first acquaintance, using a rotating frame of reference seems to make things unnecessarily complicated. However, it is well worth coming to grips with, as using a rotating frame can considerably simplify the visualization of some features of MRI, and it is used extensively. There is a program on the CD to introduce the rotating frame of reference. Start the program as indicated in Chapter 1; the program should appear as in Figure 3.12.

Let us consider a magnetization vector that has been tipped through an angle of 30° (Figure 3.13). It will precess around the direction of the main

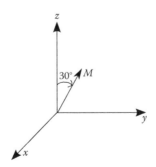

FIGURE 3.13
The net magnetization vector tipped through 30°, so that the flip angle is 30°.

field, the z direction. First we will use the rotating frame of reference program to consider what this looks like from a nonrotating frame of reference, which is often called the *laboratory frame*. There are three parameters that can be set in the program. These are labeled along the top: *Flip angle*, *View*, and *Vector frequency*. These should be set (Figure 3.12) as follows:

- Flip angle: Fixed angle
- View: Three-dimensional (3D)
- Vector frequency: Same frequency as rotating frame of reference

Under the parameter-setting area there are three white squares, which we shall refer to as frames. Click in the leftmost of the three frames, which is labeled *Nonrotating or laboratory frame of reference*. A short animation runs, in which the arrow represents the net magnetization vector. The x, y, and z axes are shown in black. The vector is precessing about the vertical z direction, and the angle between the vector and the z axis is 30°.

Now, we will view the same vector but make the axes rotate at the same rate, and in the same direction, as the vector. Note that we are still viewing the motion from the laboratory frame of reference. Leave the settings as they are and click anywhere in the middle frame, which is labeled *Rotating frame of reference seen from laboratory frame*.

In this case you will see the axes spin with the vector. These rotating axes do not seem to be helping much, but that is because we are still viewing the motion from the laboratory frame of reference. The vector, however, now has a fixed relationship to the x and y axes. This is the first clue that we really are going to make things easier by using a rotating frame of reference. Play the animation again by clicking anywhere in the middle frame. Focus on the vector and the black axis beneath it, which is labeled with a dot (Figure 3.14). The vector and axis are now moving together, because the frame of reference is rotating at the same frequency as the vector.

Question 3.3

At what frequency does the net magnetization vector precess in a uniform magnetic field?

Answer

The net magnetization vector precesses at the Larmor frequency, which is given by Equations 3.1 and 3.4.

Question 3.4

At what frequency should the rotating frame of reference rotate to make the net magnetization vector appear in a fixed location relative to the x and y axes?

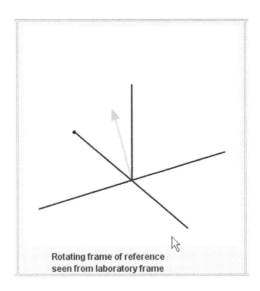

Rotating frame of reference
seen from laboratory frame

FIGURE 3.14

An image from the animation in the rotating frame of reference program that shows the rotating frame of reference viewed from the laboratory frame. The vector maintains a fixed relationship with the x and y axes.

Answer

The rotating frame of reference should rotate at the same frequency as the net magnetization vector, so it also rotates at the Larmor frequency, given by Equations 3.1 and 3.4. Rotating frames of reference always rotate at the Larmor frequency. In the rotating frame of reference program, the *Same frequency as rotating frame of reference* option for the vector frequency means that both the axes and the vector are set to rotate at the Larmor frequency. At the top right you will notice two further options for setting the vector frequency, which are grayed out at present. We will use these later.

Now imagine that instead of viewing from the laboratory frame and watching the axes spinning around, you are viewing from a location on the rotating axes. You too will be rotating, and because of this, the rotating axes will appear to be stationary. Leave the settings as they are and click anywhere in the right-hand frame, which is labeled *Rotating frame of reference seen from rotating frame.*

Do not be alarmed by the lack of motion in the right-hand frame—that is the important point about the rotating frame of reference—the rotating frame of reference removes the precessional motion. By viewing a vector from within a frame of reference that is rotating at the same frequency as the vector, the motion is stopped. This makes changes associated with differences in frequency much easier to visualize.

Question 3.5

In which of the following are all parts of the statement true?

(a) It is usual to use a rotating frame of reference to visualize a precessing vector, and the rotating frame is best viewed from within a rotating frame of reference.
(b) It is usual to use a rotating frame of reference to visualize a precessing vector, and the rotating frame is best viewed from an external stationary position.
(c) The rotating frame of reference rotates around the z axis.
(d) The rotating frame of reference rotates around the x axis.
(e) If the frequency of the rotating frame of reference matches the precessional frequency of the protons in the system, then the net magnetization vector will appear stationary within that frame of reference.

Answer

The true statements are (a), (c), and (e). One must view from within the rotating frame of reference or the motion is not removed. In the program, this difference is seen when comparing the middle animation and the animation at the right. The rotating frame of reference rotates with the Larmor frequency about the z axis, so that the xy plane appears to spin. Only if both frame of reference and vector are at the same frequency will they be stationary relative to one another.

We will now run animations that show both the net magnetization vector (arrow) and its xy component (in green). Change the *View* selection so that the parameters are set as follows:

- Flip angle: Fixed angle
- View: 3D plus xy projection
- Vector frequency: Same frequency as rotating frame of reference

Click once in each of the three frames, from left to right. You do not need to wait for the first to finish before clicking on the next; the three animations will run at once, though they will be slightly out of step with each other.

Visualization of the xy component of the magnetization vector is important because the MR signal is always measured perpendicular to the main field, in the xy plane. It is the size of the xy component of the magnetization vector that determines the signal size, and not the magnitude of the net magnetization vector. We shall see in Chapter 4 that the xy plane is important for understanding the process of T2 relaxation.

We will now run animations that show only the xy plane and the xy component of the magnetization vector (green); this is a very commonly used representation. Change the *View* selection so that the parameters are set as follows:

- Flip angle: Fixed angle
- View: *xy* plane
- Vector frequency: Same frequency as rotating frame of reference

Click in the leftmost frame, to see the *xy* component of the vector rotating in the laboratory frame of reference. Click in the middle frame, and you see the *xy* component and the rotating frame of reference both rotating at the same frequency, so that the *xy* component of the vector appears to be fixed along the *x* axis (labeled with a dot). Finally, click in the right-hand frame. Viewed from within the rotating frame of reference, the *xy* component of the magnetization vector is stationary (Figure 3.15). Note that the axes of the rotating frame of reference are labeled *x'* and *y'* instead of *x* and *y*. This labeling is conventional for the rotating frame of reference. However, because the rotating frame is used so much in MRI, the annotation is often omitted and axes labeled *x* and *y*, without primes, often represent a rotating frame of reference. In all three *xy* plane views, the direction of precession is indicated by the curved black arrow.

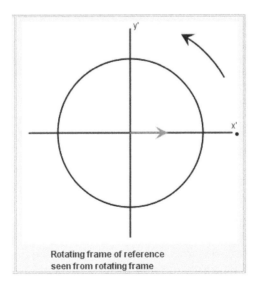

Rotating frame of reference
seen from rotating frame

FIGURE 3.15
The rotating frame of reference program showing the *xy* plane of rotating frame of reference viewed from within the rotating frame.

Question 3.6

In the diagrams in the program, the *xy* component is shown with a length 0.5 times the radius of the black circle. Why does the *xy* component of the magnetization vector have a magnitude of 0.5 in our examples?

Answer

The length of 0.5 arises because we have been using a vector with a flip angle α of 30°. By a simple geometric argument ($M_{xy} = M_0 \sin \alpha$), the magnitude of the *xy* component of the vector is half that of the vector itself. The circle shown is the maximum value that can be taken by the *xy* component (i.e., M_0), when the vector has a flip angle of 90°.

Question 3.7

Think about what would happen if the rotating frame of reference were rotating with a different frequency from the spins (and hence from the magnetization vector). If the spins were precessing at a lower frequency than the Larmor frequency of the rotating frame of reference, would you still expect to see no relative motion between the vector and the x' axis?

Answer

No, in this situation there will be relative motion between the vector and the x' axis. Recall that the reason that we see no difference in the motion is because they are rotating at the same rate. As soon as a difference is introduced, the vector and the axis will move apart. This relative motion is useful when considering situations where spins are precessing with a range of frequencies, arising from random, or deliberately introduced, small changes in the field strength. We use the rotating frame of reference in these situations in Chapters 4 and 8.

We will now use the rotating frame of reference program to see what the expected relative motion looks like. Change the *Vector frequency* selection so that the parameters are set as follows:

- Flip angle: Fixed angle
- View: *xy* plane
- Vector frequency: Vector lower frequency

Click in the leftmost frame. You will see the *xy* component of this slower vector shown in blue and rotating in the laboratory frame of reference. The vector starts aligned with the horizontal axis, and finishes pointing vertically down the vertical axis. The vector has gone three quarters of the way around in the time that a vector rotating at the Larmor frequency would complete 360° of rotation.

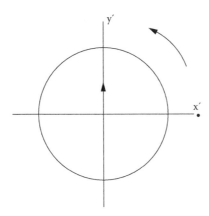

FIGURE 3.16
Illustration for Question 3.8, showing the last frame of the animation for a vector with a higher frequency than the Larmor frequency, seen from within the rotating frame of reference.

Click in the middle frame and you see the rotating frame of reference rotating anticlockwise (the direction of precession) at the Larmor frequency, and the xy component rotating more slowly (because it has a lower frequency). Had the xy component been rotating at the same frequency, it would have appeared to be fixed along the x' axis; however, because it is rotating more slowly, it gets out of phase with the axis.

Click in the right-hand frame, and, as usual, when viewed from within the rotating frame of reference, the x' and y' axes are stationary. The $x'y'$ component of the more slowly precessing magnetization vector moves away from the position that it would have held (along the x' axis) had it been rotating at the same frequency. Because the direction of precession is anticlockwise, as indicated by the curved black arrow, the slowly precessing vector appears to move clockwise relative to the axes that are rotating at the Larmor frequency. A vector rotating at the Larmor frequency would be stationary along the x' axis.

Predict what you expect to see using a vector with a higher frequency than the Larmor frequency of the rotating frame of reference. Check using the rotating frame of reference program with *Vector frequency* set to higher frequency. The faster vector appears in red.

Question 3.8

Figure 3.16 shows the appearance (when the animation has finished running) of the $x'y'$ component seen from within the rotating frame of reference. The vector in this case had a higher frequency than the Larmor frequency. Draw a copy of the diagram and add the positions of:

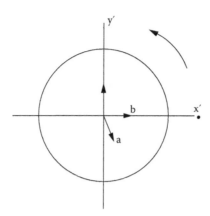

FIGURE 3.17
Feedback for Question 3.8.

  (a) The $x'y'$ component of a vector with a lower frequency than the Larmor frequency and the same flip angle

  (b) The $x'y'$ component of a vector rotating at the Larmor frequency and the same flip angle

Answer

Your drawing should look like Figure 3.17. The vector components all have the same magnitude because the flip angle is the same in each case. The vector at the Larmor frequency (labeled b) remains in position along the x' axis. The slower vector (a) lags behind, while the faster one (which was already drawn) is ahead. The curved arrow in the figure indicates that all precession is in an anticlockwise direction. It is important when using diagrams of this sort to always take note of the direction of precession. In your answer you may have drawn the slower vector aligned with the negative y' axis. This would be a vector that differs from the Larmor frequency by the same amount as the vector shown in Figure 3.16, but has a frequency that is lower than, rather than higher than the Larmor frequency. In Figure 3.17, vector (a) represents a more general case, where the slower vector differs from the Larmor frequency by a smaller amount.

Question 3.9

Draw a copy of the diagram in Figure 3.16. Add the $x'y'$ component of a vector rotating at the Larmor frequency, which has undergone a 90° flip instead of the 30° flip used in our previous examples.

Answer

The vector component in your diagram should be drawn along the x' axis because it is precessing at the Larmor frequency. Previously we have shown a vector that has been flipped by only 30°, so the $x'y'$ component had a magnitude of 0.5 (see Question 3.6). For a 90° flip, the magnitude of the vector is 1, and the tip of the vector will touch the circle in the diagram.

The supply of an RF pulse is an essential step in MRI. The longer the RF energy is applied, the larger the flip angle for the net magnetization vector. The motion of the net magnetization vector during the application of the RF pulse is usually shown in the rotating frame of reference, where the movement is an easily understood tipping away from alignment with the z axis, with no added rotation. We can use the program to see how this tipping motion would appear without the rotating frame of reference. Change the *Flip angle*, *View*, and *Vector frequency* selections so that the parameters are set as follows:

- Flip angle: Increasing flip angle
- View: 3D
- Vector frequency: Same frequency as rotating frame of reference

Click in the leftmost frame. The vector is seen to make a spiraling motion. Click in the right-hand frame for the correct rotating frame of reference view, and the complex motion is simplified. The vector tips over without any rotation. The rotating frame of reference is always used when discussing the supply of RF energy to tip the net magnetization vector, and the effect is a simple tipping, with no spiral motion.

TO CLOSE THE PROGRAM

Click on the cross at the top right of the window and then confirm that you want to end the program.

THE GRADIENT COILS IN AN MR SYSTEM

In this chapter we have seen that small differences in precessional frequency can lead to phase differences in the $x'y'$ plane. Such phase differences can be exploited for imaging, as we shall see in subsequent chapters. Differences in precessional frequency can be deliberately introduced by using gradient coils to make small differences to the value of the main field, so that some nuclei precess at slightly different rates than expected for the main field alone. MR systems have three sets of gradient coils to apply gradients in the x, y, and z directions. The gradient coils have thick windings, but only a few loops. They are built into the walls of the bore of the magnet. Gradient strengths are measured in mT m^{-1}, and a typical gradient at 1.5 T is 20 mT m^{-1}.

3.7 Chapter Summary

The basic physical principles of MRI were introduced in this chapter. We saw that the hydrogen nucleus not only is abundant in the body, but also has the property called spin. Nuclei with spin, such as the hydrogen nucleus, behave in a particular way when placed in a strong magnetic field. They precess around the direction of the field, and will absorb electromagnetic radiation, providing it has a characteristic frequency known as the Larmor frequency. To help describe and visualize this process of nuclear magnetic resonance, the net magnetization vector, which is a vector representing the magnetization from the very large number of spins, is used. The vector is tipped out of its equilibrium position when energy is supplied at the Larmor frequency. The net magnetization vector is considered in terms of two perpendicular components, the longitudinal and transverse components. It is the transverse component that gives the signal that is measured in MRI. The vector is always viewed using a rotating frame of reference, and the interactive rotating frame of reference program was used in this chapter to demonstrate why the rotating frame is helpful.

At intervals throughout the chapter the basic principles were linked to the roles of the three main components of any MR system: the main magnetic field, radio frequency coils, and gradient coils. The use of radio frequency coils will be considered further in Chapter 4. Gradient coils are discussed in Chapters 5 to 7. Scanner technology develops rapidly, and although details of the systems change, the three components are so closely related to basic principles that they are always present.

The emphasis in this chapter was on how the supply of energy affected the net magnetization vector; in the next chapter we shall concentrate on the changes that take place when the energy supply is switched off.

Reference

1. Krestel, E. 1990. *Imaging systems for medical diagnostics*, 143. Berlin: Siemens AG.

4

Relaxation Mechanisms

In the last chapter it was seen that the net magnetization vector could be moved from its equilibrium state, aligned with the z direction, by supplying energy from radio frequency (RF) radiation. Furthermore, tipping the magnetization vector means that a signal can be detected. In this chapter the emphasis is on what happens when the RF is switched off. The system will revert back to its original state, in a way that affects the measured signal in a predictable manner. The relaxation mechanisms allow us to predict tissue contrast. In this chapter the *Saturation Recovery* program is used to help the reader gain familiarity with the effects of relaxation on signal size.

4.1 Learning Outcomes

When you have worked through this chapter, you should:

- Be able to explain the physical basis for the relaxation mechanisms and define T1, T2, and T2*
- Be able to use saturation recovery graphs to predict tissue contrast

4.2 T1 and T2 Relaxation

The change back to the equilibrium state when the RF is turned off is called *saturation recovery*. Recovery is considered in terms of the two perpendicular components of the net magnetization vector, M_z and M_{xy}, as shown in Figure 4.1. When the system is in equilibrium, with an external magnetic field (B_0) applied but no radio frequency (B_1) supplied, it is said to be unsaturated. Saturation occurs following the supply of energy in the form of RF radiation, and the system is saturated when there is no net magnetization in the direction of B_0 (the z direction), when M_z is zero. Saturation recovery is the process of the change from the saturated condition to the equilibrium state.

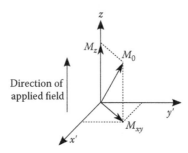

FIGURE 4.1
The components of the net magnetization vector, viewed in a rotating frame of reference.

The rate at which saturation recovery takes place is governed by the relaxation mechanisms of the material, and we shall now look at the two relaxation mechanisms involved.

4.2.1 Longitudinal Relaxation

Longitudinal relaxation is concerned with the increase in the M_z component of the net magnetization vector, M, as M returns to alignment with the z axis. There are several alternative names for the process of longitudinal relaxation: *T1 relaxation, spin-lattice relaxation,* or *thermal relaxation*. During longitudinal relaxation, energy is transferred from the excited nucleus to its surroundings, which are described as the lattice, but not to the nucleus itself. It is important to recognize that longitudinal relaxation occurs only after a stimulated change to a higher energy state, achieved by the RF pulse. The change in energy state cannot be spontaneous.

The shape of the curve associated with longitudinal (T1) relaxation is shown in Figure 4.2. Note that the start of M_z recovery is conventionally plotted at a time immediately after a 90° RF pulse. The 90° RF pulse tips the net magnetization vector into the $x'y'$ plane, so there is no longitudinal

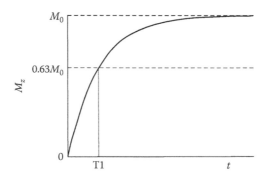

FIGURE 4.2
The recovery of longitudinal magnetization (M_z) and its relation with the value of T1.

component, so $M_z = 0$ at time $t = 0$. The recovery of the M_z component is described by a function that includes an exponential decay term, with a time constant T1:

$$M_z = M_0 \left[1 - \exp\left(-\frac{t}{T1} \right) \right]$$

(4.1)

Question 4.1

If necessary, refer back to Chapter 2 to remind yourself about exponential functions, and then select the true statements from this list:

(a) In a plot of M_z against t, T1 is the time taken for the system to become unsaturated again.

(b) In a plot of M_z against t, T1 is the time taken for the value of M_z to reach 63% of its final value.

(c) In a plot of M_z against t, T1 is the time taken for the value of M_z to reach 37% of its final value.

(d) If the time constant T1 were longer than that in the example shown in Figure 4.2, the curve would reach the value M_0 at a smaller value of t than in the figure.

(e) If the time constant T1 were longer than that in the example shown in Figure 4.2, the curve would reach the value M_0 at a greater value of t than in the figure.

Answer

You should have decided that the true answers were (b) and (e), and that the others were false. If you thought that (c) was correct and (b) was not, this might be because the presence of the expression for exponential decay in Equation 4.1 led you to choose the answer containing 37%. Note, however, that the full expression includes a term where the exponential decay is subtracted from 1, and so the percentage recovery is 63%. When choosing between (d) and (e), note that a longer time constant means that it takes longer for the recovery to take place, so the curve will reach the value M_0 further to the right, and at a higher value of t. Similarly, a more rapid recovery is associated with a smaller value for T1 (Figure 4.3).

As we can see in Figure 4.3, after an elapsed time of T1 ms, only partial recovery has taken place. A longer period, of about five times the T1 value, is needed for full recovery to take place so that the system is unsaturated again. Armed with this knowledge, if presented with a T1 recovery curve for a material, you would be able to estimate the value of T1. We shall see later how this value lets you determine if the material will appear bright or dark in an image. Use Equation 4.1 to calculate the time period, in multiples of T1, for 95%, 99%, and 99.9% recovery to take place. Ensure that

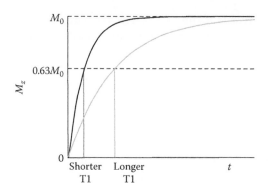

FIGURE 4.3
Comparison of the longitudinal recovery curves for two materials with different values of T1.

your answers agree with the rule of thumb that about five T1s are needed for full recovery.

Question 4.2

Estimate the values of T1 for each of the two materials whose T1 recovery curves are plotted in Figure 4.4.

Answer

The T1 values are estimated by reading from the graph the value of t that corresponds with $M_z = 0.63M_0$. You should have estimates of about 250 ms and 800 ms. These values are typical of the values for tissues in the body, and we shall now go on to consider the factors that determine the size of T1 in a material.

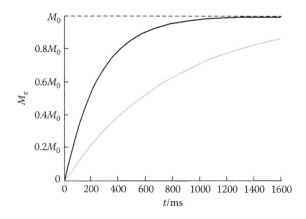

FIGURE 4.4
Graph for Question 4.2 showing longitudinal recovery curves for two materials with different T1 values.

4.2.1.1 Typical Longitudinal Relaxation Times

Molecules in tissue undergo constant random motion, in which their positions and orientations change continually. This motion is known as molecular tumbling. The tumbling rate depends on the particular molecules involved, and it is slower for large molecules. The molecules of the lattice (i.e., the surroundings of the excited nucleus) are tumbling, and as a result, their magnetic fields fluctuate at a rate that depends on the tumbling frequency. If the magnetic fields fluctuate at a frequency near the Larmor frequency, then conditions are right for the energy exchange to take place between the excited nucleus and the lattice. This leads to shorter values of T1. When the molecules of the lattice are rotating at a rate different from the Larmor frequency (both much slower and much faster), then the value for T1 is greater than at the Larmor frequency (Figure 4.5).

From Figure 4.5 it can be seen that:

- T1 is short, giving rapid relaxation, for tissue in which water molecules are partially bound to large molecules such as proteins, where molecules are rotating at a rate near to the Larmor frequency.

- T1 is longer, giving slow relaxation, for both solids and liquids where molecules are likely to be rotating at rates either lower or higher than the Larmor frequency.

As a result of these variations, different tissues have characteristic T1 values. The characteristic T1 values are important properties for imaging, as they can be used to generate image contrast. In general, in tissue T1 increases with increasing water content and decreases with increasing

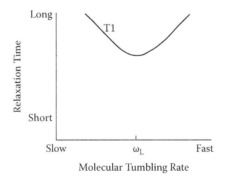

FIGURE 4.5
Variation of T1 with the molecular tumbling rate, where ω_L represents the Larmor frequency. The molecular tumbling rate is slow for solids, large molecules, and bound water, and fast for liquids, small molecules, and free water. (Adapted from Krestel, E., *Imaging Systems for Medical Diagnostics*, Siemens AG, Berlin, 1990. With permission.)

macromolecular (e.g., proteins, lipids) content, so that fat, for example, has a short T1.

Question 4.3

You are proofreading a draft written by a colleague, in which it is stated that the T1 values at 1.5 T for cerebrospinal fluid and fat are 250 ms and 2,500 ms, respectively. How would you comment on this part of the draft?

Answer

You should suspect that the two T1 values have been reversed, because fat has a short T1, and high-water-content materials like cerebrospinal fluid have long values of T1.

Question 4.4

Curves for the value of M_z following a 90° RF pulse are plotted for two tissues, tissue A and tissue B. At $t = 500$ ms, the curve for tissue A has a value of $0.63M_0$. The curve for tissue B has a value of $0.63M_0$, occurring 250 ms after the curve for tissue A has that value. What are the T1 values for the two tissues?

Answer

T1 for tissue A is 500 ms, and T1 for tissue B is 750 ms. These values are obtained because T1 for each tissue is the time taken for M_z to reach a value of $0.63M_0$.

4.2.2 Transverse Relaxation

Transverse relaxation represents the decay of the net magnetization vector (M) in the $x'y'$ plane as it returns to alignment with the z axis, and so is considered in terms of the decay of the transverse component M_{xy} of the magnetization vector. The alternative names for the process of transverse relaxation are *T2 relaxation* and *spin-spin relaxation*. During transverse relaxation there is no energy transfer, unlike the situation for longitudinal relaxation. Instead, there is interaction between the magnetic moments of neighboring protons, which means that the size of M_{xy} is reduced. Each proton in a material is in a very slightly different chemical environment, and so it experiences a slightly different magnetic field from those experienced by other protons. As the field size dictates the frequency of precession of the spin, this means that there will be a loss of phase coherence between the spins as they spin at different rates. The vector sum of the contributions of all the spins to M_{xy} gets smaller as the spins get out of phase with each other. This is illustrated in Figure 4.6, which shows the $x'y'$ plane in the rotating frame of reference. At time (a), before the application of the 90° RF pulse, the value of M_{xy} is zero and the spins are precessing

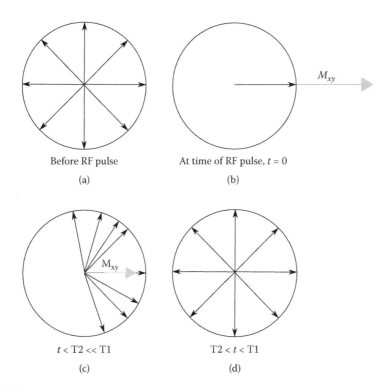

FIGURE 4.6

Illustration of the dephasing of spins in the $x'y'$ plane of the rotating frame of reference, which is associated with T2 relaxation. (a) Before the RF pulse the spins are out of phase and M_{xy} is zero. (b) Immediately after the RF pulse, the spins are all in phase, leading to a large resultant M_{xy} component of the magnetization vector. (c) As time goes by, dephasing leads to the decay of M_{xy} with a time constant T2. (d) M_{xy} is zero again, as it was before the RF pulse, at a time less than the T1 relaxation time.

out of phase. When the RF pulse is applied at (b), it brings all the spins into phase, and the vector sum for M_{xy} is large. As time passes (c), the differing chemical environments for the spins mean that they again get out of phase, and the vector sum for M_{xy} is reduced in size. After a period of time smaller than T1 for the material (d), the spins are completely out of phase again and M_{xy} is zero.

DIFFERENT FREQUENCIES IN THE ROTATING FRAME OF REFERENCE

Refer back to the section of the rotating frame of reference to remind yourself how spins precessing at rates that differ from the Larmor frequency appear to pull away from each other in the $x'y'$ plane in the rotating frame of reference.

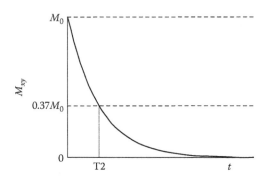

FIGURE 4.7
The decay of transverse magnetization (M_{xy}) and its relation with the value of T2.

A perfectly homogeneous B_0 field is assumed when considering the T2 relaxation process, so that the only variations in magnetic fields are due to the different chemical environments of neighboring spins. Dephasing arising from these spin-to-spin differences is sometimes called *intrinsic* dephasing. The consequences of inhomogeneities in the B_0 field are discussed in Section 4.2.4.

The shape of the curve for the decay of the M_{xy} component is shown in Figure 4.7. The start of M_{xy} decay is conventionally plotted at a time immediately after a 90° RF pulse. The 90° RF pulse that brings the spins into phase has the effect of tipping the net magnetization vector into the $x'y'$ plane, so $M_{xy} = M_0$ at time $t = 0$. The decay of the M_{xy} component is described by an exponential decay with time constant T2:

$$M_{xy} = M_0 \left[\exp \left(-\frac{t}{T2} \right) \right] \tag{4.2}$$

MNEMONICS FOR T1 AND T2

T1 is the longitudinal relaxation time. This can be remembered by noting the similarity between 1 and the *l* of longitudinal. Furthermore, energy is transferred to the lattice, which is another word beginning with *l*.

T2 is the transverse relaxation time. This can be remembered by noting that T2 involves two occurrences of the letter *T*, in the words *transverse* and *time*.

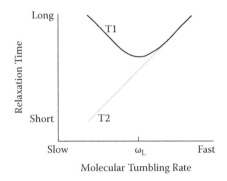

FIGURE 4.8
Variation of T2 with the molecular tumbling rate, where ω_L represents the Larmor frequency. The molecular tumbling rate is slow for solids, large molecules, and bound water, and fast for liquids, small molecules, and free water. (Adapted from Krestel, E., *Imaging Systems for Medical Diagnostics*, Siemens AG, Berlin, 1990. With permission.)

4.2.2.1 Typical Transverse Relaxation Times

Like T1 relaxation, T2 relaxation is associated with the molecular tumbling rate, but for T2 the association with the Larmor frequency is less strong. Spin-spin interactions are more likely to take place when the molecules are moving slowly compared with the Larmor frequency. The slow changes in magnetic field cause dephasing, and so in this situation the T2 relaxation time is short. In contrast, rapidly tumbling molecules have such rapid changes in magnetic field that their average effect is zero and dephasing of spins occurs only slowly, so T2 times are long (Figure 4.8).
In Figure 4.8 it can be seen that:

- T2 is short, giving rapid relaxation, for molecules that move slowly, such as proteins and lipids.
- T2 is longer, giving slow relaxation, for molecules with higher molecular tumbling rates, such as free water molecules.

As a result of these variations, and as is the case for T1, different tissues have characteristic T2 values. In general, in tissue, T2 (like T1) is high for water-based tissue and lower for fat-based tissue.

Question 4.5

Select the true statements:

(a) When T1 has a high value, M_z increases more rapidly than when T1 has a low value.
(b) When T2 is long, M_{xy} decreases to zero over a longer period of time than when T2 is short.

(c) When T2 has a high value, M_{xy} decreases to zero over a longer period of time than when T2 is low.

(d) When T1 is short, M_z increases more rapidly than when T1 is long.

(e) When T1 has a low value, M_z increases more rapidly than when T1 is high valued.

Answer

The statements in Question 4.5 used a variety of different ways to describe the size of the relaxation times and their relation to saturation recovery. All the statements, except (a), are true. If you got confused, remember that short relaxation times, or small values of the relaxation time, are associated with rapid, or fast, recovery. Long relaxation times, or high values of the relaxation time, go with long, or slow, recovery.

4.2.3 Comparison of T1 and T2 Relaxation Times

It was seen in Figure 4.5 that the most rapid T1 relaxation requires the molecular tumbling rate to be close to the Larmor frequency, and any other frequency results in increased relaxation time. Rapid T1 relaxation will be possible only for a limited number of the molecules present. In contrast, rapid T2 relaxation is possible for a whole range of molecules. As a result, T1 values are longer than T2 values. In tissue, T1 is between 5 and 10 times longer than T2. The value of T2 is always smaller than the value of T1 for a material. This relationship is consistent with the vector description seen in Figure 4.1: as long as the M_{xy} component is greater than zero, the M_z component cannot have attained the value (M_0) that is associated with complete recovery.

Typical values for T1 and T2 for healthy human tissue are shown in Table 4.1.

Question 4.6

On a student's Web site it is stated that in some practical experiments using foodstuffs, the T2 values for a block of butter and orange juice could not be differentiated. Which of the two would you expect to be higher? Noting the results shown in Table 4.1, suggest why the results of the experiment were not able to show a difference conclusively.

Answer

You should have noted that the T2 value for orange juice is expected to be higher because it is a liquid, while butter, which is fatty, would have a short T2. In Table 4.1 it can be seen that measurements of T2 give large error ranges, some over 40%. The errors arise from both experimental limitations and the natural variability between

TABLE 4.1

Typical Values for Relaxation Times of Human Tissue at 1.5 T

Tissue	T1/ms	T2/ms
Brain		
Cerebrospinal fluid (CSF)[a]	800–20,000	110–2,000
Gray brain matter[b]	920 ± 160	101 ± 13
White brain matter[b]	790 ± 130	92 ± 22
Fat[b]	260 ± 70	84 ± 36
Kidney[b]	650 ± 180	58 ± 24
Liver[b]	490 ± 110	43 ± 14
Skeletal muscle[b]	870 ± 160	47 ± 13

[a] Large ranges because of the influence of CSF flow. From Fletcher, L. M. et al., *Magn. Reson. Med.*, 29, 623, 1993. Copyright © 1993 by Williams & Wilkins. Reprinted with permission of Wiley-Liss, Inc., a subsidiary of John Wiley & Sons, Inc.

[b] From Bottomley, P. A. et al., *Med. Phys.*, 11, 425, 1984. (With permission of the American Association of Physicists in Medicine.)

biological samples. It is possible that the student's experiments had similar large errors associated with them, and for a small number of measurements, statistically significant differences would not be measured.

Question 4.7

M_{xy} relaxation curves following a 90° RF pulse are plotted for two tissues, C and D. At $t = 80$ ms, the curve for tissue C has a value $M_{xy} = 0.37M_0$. Tissue D has a value of T2, which is 20 ms longer than T2 for tissue C. What is the T2 value for tissue C? At what value of t is $M_{xy} = 0.37M_0$ for tissue D?

Answer

T2 for tissue C is 80 ms. $M_{xy} = 0.37M_0$ for tissue D when $t = 100$ ms. These values are inferred because T2 for each tissue is the time taken for M_{xy} to reach a value of $0.37M_0$.

Question 4.8

An erratum slip is enclosed with a radiology textbook. It starts: "In Table 25 on page 768, the values for T1 and T2 for adipose tissue are given as 70 ms and 200 ms, respectively." Suggest what the correction might be.

Answer

The point to pick up on here was that in the printed version of the textbook the value of T1 was less than the value for T2, which cannot be the case. It is possible that the values were simply transposed, but they might have been misprinted.

Question 4.9

Select the statements where all the assertions in the statement are true:

(a) T2 is the transverse relaxation time, and it is associated with the recovery of the M_0 component of the magnetization vector.
(b) The longitudinal relaxation time is also called the spin-spin relaxation time and is associated with dephasing of the spins in the $x'y'$ plane.
(c) The spin-lattice relaxation time is associated with recovery of the M_z component of the magnetization vector; it is longer than the relaxation time associated with the M_{xy} component of the magnetization vector.
(d) The more rapid the dephasing of spins in the $x'y'$ plane, the shorter the spin-spin relaxation time.
(e) Spin-spin relaxation is always more rapid than transverse relaxation, and it arises from the dephasing of spins resulting in the decay of the signal in the $x'y'$ plane.

Answer

The statements in Question 4.9 each included several assertions, and in each statement there was a true assertion; however, only in statements (c) and (d) were all the assertions true.

4.2.4 T2*

It was noted earlier that a perfectly homogeneous B_0 field is assumed when considering the T2 relaxation process. If B_0 is not perfectly homogeneous (which it never is), then additional dephasing of the spins arises from the inhomogeneities in the magnetic field.

- The time constant for the decay that results from both intrinsic dephasing and dephasing from field inhomogeneities is T2* (this is pronounced as T2 star).
- T2* is shorter than T2 (Figure 4.9).

The inhomogeneities in B_0 can arise from:

- Inherent defects in the magnet itself
- Susceptibility-induced field distortions from tissue or other materials in the field

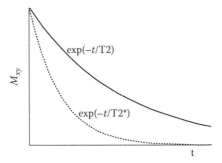

FIGURE 4.9
A comparison of T2 decay and the more rapid T2* decay arising from magnet inhomogeneity.

We saw in Chapter 3 that the M_{xy} component of magnetization always gives rise to a signal known as the free induction decay (FID) signal. Now that we have learned how relaxation processes lead to the decay of the M_{xy} component, we can see that the FID signal must also decay with a time constant T2* as shown in Figure 3.11.

4.2.5 Relaxation Rates, R1, R2, and R2*

The relaxation rate is the reciprocal of relaxation time, for example, R2 = 1/T2. Relaxation rates may be added together to combine the effects of various relaxation times. For example, to express the effect of the relaxation time due to inhomogeneities, $T2_{inh}$, in terms of T2 and T2* the following equation applies:

$$\frac{1}{T2^*} = \frac{1}{T2} + \frac{1}{T2_{inh}} \tag{4.3}$$

Worked Example

T2 for a sample is known to be 60 ms. When T2* is measured it has a value of 40 ± 8 ms. What is $T2_{inh}$ in this case?

Rearranging Equation 4.3 gives

$$\frac{1}{T2_{inh}} = \frac{1}{T2^*} - \frac{1}{T2}$$

Substituting the known values and working in ms,

$$\frac{1}{T2_{inh}} = \frac{1}{40} - \frac{1}{60}$$

Rearranging,

$$T2_{inh} = 120 \text{ ms}$$

Question 4.10

Select the one true statement from the following three statements:

(a) If the B_0 field is perfectly homogeneous, then there is no decay of the $x'y'$ component of magnetization due to dephasing.
(b) If the B_0 field is perfectly homogeneous, then the $x'y'$ component of magnetization decays with a time constant T2.
(c) If the B_0 field is perfectly homogeneous, then the $x'y'$ component of magnetization decays with a time constant T2*.

Answer

The true statement is (b).

Question 4.11

Select the one true statement from the following three statements:

(a) Intrinsic dephasing arises because spins are in differing molecular environments.
(b) Intrinsic dephasing arises because of variations in the magnetic field caused by the presence of a patient.
(c) Intrinsic dephasing arises because the main magnetic field is inhomogeneous.

Answer

The true statement is (a).

Question 4.12

Select the one true statement from the following three:

(a) T2* is the time constant associated with dephasing arising solely from the different molecular environments experienced by spins.
(b) T2* is the time constant associated with dephasing arising solely from defects in the main magnetic field.
(c) T2* is the time constant associated with dephasing from the different molecular environments experienced by spins, defects in the main magnetic field, and the effects of susceptibility.

Answer

The true statement is (c). For all three questions, remember that T2 is associated with intrinsic dephasing, and T2* additionally is affected by inhomogeneities in the field that can arise from two sources: defects in the main magnetic field and the effects of susceptibility.

4.3 Effect of Magnetic Field Strength on Relaxation Mechanisms

Each MR system has a fixed magnetic field strength, but the fixed field strength is not the same for all systems. The field strength of the magnetic field of the chosen system affects the relaxation properties of tissue. A sketch of the variation of T1 and T2 with field strength is shown in Figure 4.10.

T1 relaxation times increase with increasing field strength (Figure 4.10). The increase arises because of the increased Larmor frequency associated with increased field strength (Equation 3.3). There is a corresponding reduction in the proportion of molecules in a given tissue that are tumbling close to the new Larmor frequency. T1 values are approximately doubled between 0.25 T and 1.5 T.

Unlike T1 values, at the field strengths used clinically, T2 relaxation times are not changed by increased field strength (Figure 4.10).

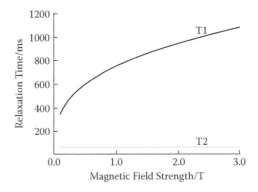

FIGURE 4.10
Illustrative graph to indicate a typical variation, for a single material, of T1 and T2 with main field magnetic field strength B_0.

4.4 Saturation Recovery Graphs and Tissue Contrast

We saw in Chapter 3 that the M_{xy} component of magnetization always gives rise to a signal (the free induction decay signal). It is common to get an idea of the relative sizes of signals expected from different tissues, after various lengths of time for recovery, by drawing simple sketch graphs, in the style of the one shown in Figure 4.11. The graphs are plotted for a standard sequence of RF pulses consisting of a 90° RF pulse at time $t = 0$, a period of recovery, and then a second 90° RF pulse, possibly taking place before full recovery has occurred. The sketch shows the M_z recovery associated with T1 relaxation after the first 90° RF pulse, and the decay of the signal-related M_{xy} component associated with T2 relaxation after the second RF pulse. To sketch this sort of graph by hand, it is not necessary to know absolute values of T1 and T2, just which one is longer.

In Figure 4.11 note how:

- The first part of the plot, up to the vertical solid line, shows the z component of magnetization, M_z, recovering after a 90° RF pulse.
  - The dotted line continuing this curve shows the recovery that would have taken place if a second 90° RF pulse were not applied.
- The second part of the plot shows the $x'y'$ component of magnetization, M_{xy}, decaying after the application of the second 90° RF

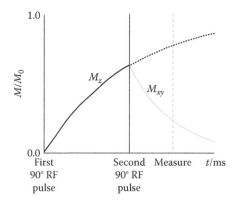

FIGURE 4.11

An example saturation recovery sketch graph to illustrate the key features of graphs of this kind, which are used to predict tissue contrast. The plot shows M_z to the left of the vertical solid line and M_{xy} to the right of it. In this sketch they have been differentiated by using black for M_z and gray for M_{xy}. A 90° RF pulse is applied at time zero, and a second 90° RF pulse corresponds with the vertical solid line. Having drawn the curve representing the first material, it is straightforward to add a curve for a material known to have shorter or longer relaxation times.

pulse. The signal in the $x'y'$ plane can be measured at any time in the second part of the plot.

- Note that the starting value of magnetization tipped into the $x'y'$ plane by the second 90° RF pulse depends on how much longitudinal recovery has taken place after the first 90° RF pulse.

4.4.1 The Saturation Recovery Program

Working with graphs of this sort will help to consolidate the information about relaxation mechanisms that you have learned in this chapter. We shall start by using a program that will plot saturation recovery graphs for you. Once you are familiar with the principles, you will probably find it easy to sketch the plots by hand. Start the Saturation Recovery program as indicated in Chapter 1. The display should look like Figure 4.12. If you now select *Tissue A (Black)* by clicking in the box next to the words

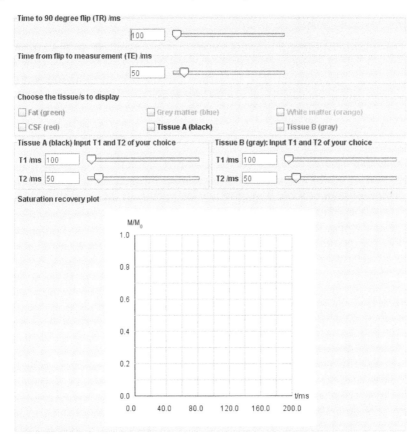

FIGURE 4.12
The graphical user interface for the Saturation Recovery program.

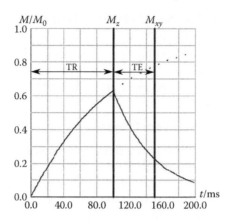

FIGURE 4.13

A saturation recovery plot generated by the Saturation Recovery program for a material with T1 = 100 ms and T2 = 50 ms. The time periods TR and TE are indicated in the figure, but do not appear when the program is run.

in the middle column, a graph appears in the lower half of the interface (Figure 4.13). The plot returned by the saturation recovery program follows the conventions outlined regarding Figure 4.11, and illustrates relaxation starting immediately after a 90° RF pulse.

- At time $t = 0$, a 90° RF pulse is applied to tip the net magnetization vector from alignment with the z axis into the $x'y'$ plane.

- In the first part of the plot, between $t = 0$ and the first vertical black line, the vertical axis indicates the value of M_z, which is the z component of the net magnetization vector. In this case the vertical black line is at $t = 100$ ms.

- At $t = $ TR (the first vertical black line), another 90° RF pulse is applied that again tips the z component of the magnetization vector into the $x'y'$ plane. Note how the amount of magnetization available to this second RF pulse is dependent on the amount of relaxation that has taken place after the first RF pulse. The reason for calling this time period TR, which stands for *repetition time*, will be explained in Chapter 8.

- In the part of the plot after the first vertical black line, the vertical axis indicates the value of M_{xy} (instead of M_z). M_{xy} is chosen rather than M_z because when an MR signal is measured, the signal is measured in the $x'y'$ plane. The z component of magnetization is not measured.

- The value of M_{xy} at $t = $ TR is the same as the value of M_z immediately before the 90° RF pulse, because a 90° RF pulse tips all the available z magnetization into the $x'y'$ plane. We have previously called this the starting value of magnetization.

- The component M_{xy} then decays. The signal is measured at the time of the second vertical black line.

- The time period between the second 90° RF pulse and measurement is called TE, which stands for *time to echo*. As for the name TR, the reason for this naming convention is explained in Chapter 8.

- You can also use the second vertical black line as a reference for considering the effect of using shorter or longer TEs.

- The curve in Figure 4.13 is for TR = 100 ms, TE = 50 ms, and a material that has T1 = 100 ms and T2 = 50 ms.

- The dotted line that continues the first curve is there to emphasize the course of T1 recovery that would take place if the second 90° RF pulse were not applied.

Question 4.13

Choose the correct statement from these two alternatives:

(a) A system is said to be saturated when there is no net magnetization in the z direction.

(b) A system is said to be saturated when there is maximum net magnetization in the z direction.

Answer

This question was included to remind you that a system is saturated when there is no net magnetization in the z direction, which is the situation immediately after a 90° RF pulse to tip the magnetization vector into the $x'y'$ plane. Saturation recovery is the process of recovery from the state of saturation, in which M_z increases from zero.

Question 4.14

(a) Is the component of magnetization shown in the first part of the plot, from $t = 0$ to $t = $ TR, M_z or M_{xy}?

(b) Is the component of magnetization shown in the second part of the plot, for $t > $ TR, M_z or M_{xy}?

(c) Is the component of magnetization that gives rise to a signal that can be detected in an MR system M_z or M_{xy}?

Answer

The statements in Question 4.14 should have reminded you about the key features of the graphs plotted using the saturation recovery program. The component of magnetization shown in the first part of the plot, from $t = 0$ to $t = $ TR, is M_z; the component of magnetization shown in the second part of the plot, for $t > $ TR, is M_{xy}; and the component of magnetization whose signal can be detected in an MR system is M_{xy}.

Now, use the program to generate some new curves. Put ticks in the boxes in the *Choose the tissue/s to display* area of the interface for fat, cerebrospinal fluid (CSF), gray matter, and white matter. Remove the tick for *Tissue A* by clicking on it. Leave the other settings as they are. Curves will appear for the selected tissues in the colors indicated beside the tick boxes. The display should appear as in Figure 4.14.

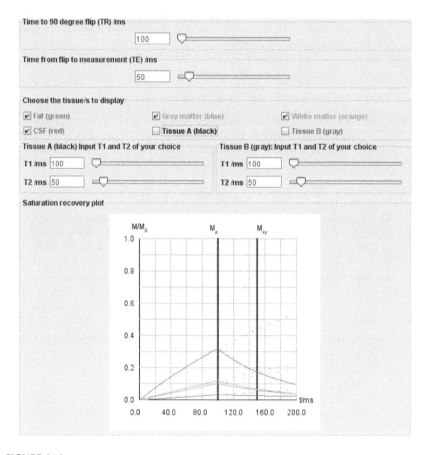

FIGURE 4.14
The graphical user interface for the Saturation Recovery program when the four named built-in tissues are selected. Use the Saturation Recovery program to see the plot in color.

Question 4.15

Use the plot that you have just generated to answer the following questions:

(a) Which has the longer T1 relaxation time, fat or CSF?
(b) Which has the longer T1 relaxation time, white matter or gray matter?
(c) If an image were acquired at the time indicated by the second vertical black line, which of the four tissues would appear brightest in the image?

Answer

The first section of the plot is the part to concentrate on for Question 4.15a, because it shows M_z, and thus indicates T1 (longitudinal or spin-lattice) relaxation. The slope of the CSF curve in the first section of the plot is less steep than the slope of the fat curve, indicating the longer T1 relaxation time for CSF. Similarly, for Question 4.15b the T1 relaxation times for white matter and gray matter must be very similar, as the orange and blue curves have a similar slope in the first section of the plot. The blue curve is slightly below the orange one, which suggests a slightly longer T1 relaxation time for gray matter. For Question 4.15c look at the second section of the plot, which shows M_{xy} and indicates T2 (transverse or spin-spin) relaxation. M_{xy} is the component that is measured as the MR signal, so the highest value leads to the brightest pixels in an image. In this case, fat (green) would appear the brightest of the four tissues.

The values for T1 and T2 that have been set in the saturation recovery program are shown in Table 4.2. The values shown in the table confirm that T1 is longer for CSF than for fat, and longer for gray matter than for white matter. You will find the tabulated values useful in later questions.

You can use different values for the time to 90° flip (TR) and the time from flip to measurement (TE) by adjusting the sliders. Try this now. If

TABLE 4.2

Values Used for T1 and T2 in the Saturation Recovery Program

	T1/ms	T2/ms
Fat	260	84
Gray matter	920	100
White matter	780	90
CSF	3,000	300

you type values into the box instead of using the sliders, you must hit the enter key to make the value register.

Question 4.16

Adjust TE and TR so that the highest signal from the four tissues comes from a tissue other than fat. If you are unsure about how to proceed, read the hint in the footnote.*

Answer

To obtain a high signal from a tissue other than fat, you should have increased the values of TR and TE to, for example, TR = 1,000 ms and TE = 250 ms, to give a plot as shown in Figure 4.15. The longer TR value means that more longitudinal relaxation has taken place for all the tissues. For T1 tissues with long T1 values, using a long TR means that much higher amounts of magnetization are now available at the time of the second 90° RF pulse than if a short TR were used. The longer TE means that there is a longer time interval between the second RF pulse and the measurement, in which T2 decay proceeds. The T2 values for the tissues other than CSF are short, so their signals decay quickly. The result is that at the second vertical line the signal for the red CSF curve is the highest.

Next, add a tick next to *Tissue A (black)*. Then use the area under the tick boxes on the left, where it says *Tissue A (black) Input T1 and T2 of your choice*, to choose your own values for T1 and T2 for tissue A. The relevant curve will be plotted in black. Toggle the ticks to remove the four tissue plots if you wish.

Most lesions, for example, a tumor, or a plaque in multiple sclerosis, have a longer T1 than the tissue in which they occur. For example, consider white matter (the orange curve) for which T1 = 780 ms and T2 = 90 ms (Table 4.2). Investigate the effect of setting the values for the black curve, which you are now using to represent the lesion, to have a T1 longer than 780 ms, perhaps something in the range of 900 to 1,100 ms. You may need to adjust TR and TE too.

---

* Hint: There are two factors that led to a high signal from fat in the initial setup, where TR = 100 ms and TE = 50 ms. First, at the time of the second 90° RF pulse, for fat a larger amount of magnetization is available to be tipped into the plane than for the other tissues because of fat's short T1. If TR were longer, even the long T1 materials will have recovered sufficiently for there to be a large component available to flip. Second, the time period before measurement is short, so differences in the T2 relaxation times between the tissues do not have time to manifest themselves. If TE were longer, it is possible that the material with the longest T2 would take so long to relax that its signal would be the highest, because the other tissues would have completed relaxation.

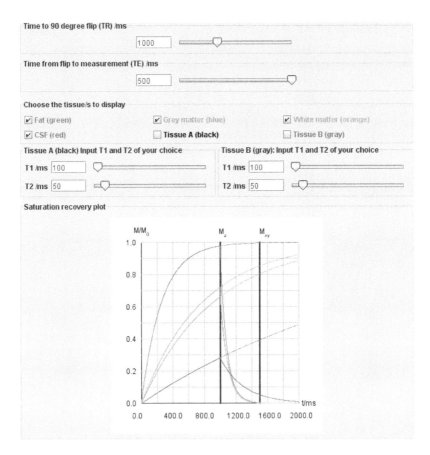

FIGURE 4.15
The graphical user interface for the Saturation Recovery program set up for a solution to
Question 4.16.

Question 4.17

Are differences in T1 emphasized by using a long or a short TR? Find
a combination of TR and TE in which the lesion (with T1 longer than
780 ms) gives a higher signal than both white matter and CSF. You
may adjust T2 of the lesion too.

Answer

Differences in T1 are emphasized by using a short TR. An example
solution for Question 4.17 is shown in Figure 4.16. If a long TR were
used, both tissues would have time to undergo full longitudinal
relaxation, so that the same amount of magnetization would be available in the z direction for the second 90° RF pulse. Any difference in
signal would be entirely due to any difference in the T2 value. In the
example in Figure 4.16, TR = 500 ms, TE = 150 ms, and the black tissue

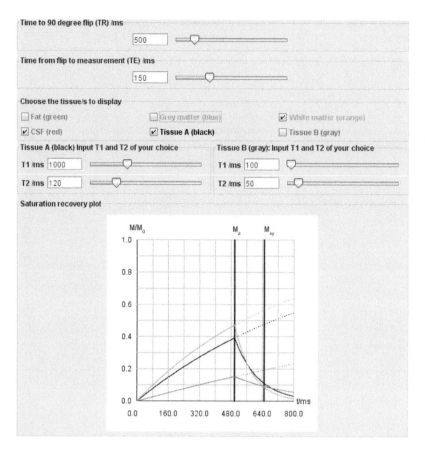

FIGURE 4.16
The graphical user interface for the Saturation Recovery program set up for a solution to
Question 4.17.

representing the lesion has T1 = 1,000 ms and T2 = 120 ms. Note that
the boxes for fat and gray matter have been deselected on the control
window, so that only the relevant curves are displayed to make the
comparisons among the lesion, white matter, and CSF.

At the TE indicated by the vertical line, the black lesion line is above
both the orange line and the red line, indicating that the increased
T1 and T2 values in the tumor will give a stronger signal than white
matter and CSF. At longer TE selections, CSF would give the strongest
signal, while at very short TE, the strongest signal of the three is from
white matter.

Switch the fat and gray matter back on, and you will see that at
short TE the fat signal is the strongest. However, because the object
of this exercise was to distinguish the increased T1 lesion tissue from
the surrounding white matter, the signal from fat was not important
here.

Question 4.18

Generate the plot in Figure 4.16 using the values TR = 500 ms, TE = 150 ms. Switch on white matter and CSF, and set black lesion tissue to have T1 = 1,000 ms and T2 = 120 ms. Keeping TR at 500 ms, at what value of TE is the signal from the lesion the same as that from white matter? What would be seen if a TE shorter than this were used?

Answer

The signals from two tissues are the same where their curves cross, in this case where TE is about 60 ms. Shorter values for TE would give a higher signal for white matter than for the lesion.

Question 4.19

Set up the Saturation Recovery program with TR = 1,500 ms and TE = 500 ms. Deselect all the tissues except gray matter and Tissue A (black). Set up T1 and T2 for the black curve to 920 ms and 50 ms, respectively. The black curve has been set to represent a tissue with the same T1 as gray matter but with a shorter T2. We know that inhomogeneities in the magnetic field lead to shorter T2 relaxation times, so what implications can be drawn from the plot about signal size in the presence of inhomogeneities?

Answer

The assumption that the main magnetic field is perfectly homogeneous means that T2 relaxation times are not shortened to T2* by the field inhomogeneities. By setting T1 and T2 for the black curve to 920 ms and 50 ms, respectively, we have arranged for the T1 to be the same as that for gray matter, but for the T2 to be shorter, as would be the case in the presence of inhomogeneities. The long TR was chosen to make a large signal available to be tipped into the $x'y'$ plane by the second 90° RF pulse. The long TE used allows the T2 decay curves to be seen in full. It can be seen in this plot that the effect of inhomogeneities is to reduce the size of the signal, as the black curve (simulating T2*) is below the blue one.

In the Saturation Recovery program, *Input T1 and T2 of your choice* will allow you to set the sliders to any value between 1 and 3,010 ms for T1, and between 1 and 500 ms for T2. Set T1 = 100 ms and T2 = 200 ms. As you know that T2 is always shorter than T1, you should have felt uncomfortable doing that! The program will give a message confirming that this is not a valid combination. Click OK to remove the message, type a smaller value into the T2 box and hit return. Although you also know that T1 is 5 to 10 times longer than T2, the program will not return an error if T1 values greater than T2 but less than five times T2 are used.

TO CLOSE THE PROGRAM

Click on the cross at the top right of the window, and then confirm that you want to end the program.

4.5 Contrast Agents

Contrast agents help to differentiate different regions of tissue by making the signal higher or lower than it would otherwise have been. Magnetic resonance imaging contrast agents work by affecting the relaxation times in the tissue in which they are taken up. Contrast agents are discussed in Chapter 9.

4.6 Chapter Summary

In this chapter the relaxation mechanisms in tissue were introduced. You learned that there are two separate processes involved with the recovery of a tissue from saturation. T1 relaxation is associated with the recovery of the longitudinal component of magnetization, and T2 relaxation with the decay of the transverse component of magnetization, which is the component that gives the MRI signal that is measured. T2* is the faster T2 decay that arises because of inhomogeneities in the main magnetic field. It was noted that T1 is always longer than T2. By considering saturation recovery after 90° RF pulses, we have seen that the measured signal strength has dependence on both T1 and T2. We have begun to consider, using saturation recovery graphs, how differences in T1 and T2 lead to contrast between different tissues.

At present, however, we have not considered how to determine from where in the subject the signal is coming, and knowledge of spatial localization is essential in order to generate an image. In the next three chapters we shall consider spatial localization, and then in Chapter 8 we will combine that knowledge with the content of this chapter to explain the production of an image with predictable contrast properties. In Chapter 9, when we consider T1- and T2-weighting, we return to the saturation recovery graphs used in this chapter, and there is an opportunity to draw the graphs by hand.

5

Slice Selection

In the previous chapter it was clear that there is considerable potential to harness the relaxation properties of tissues to generate contrast between tissues in an image. The next important consideration relevant to image generation is how to match the acquired signal to a location within the subject, which is a process termed spatial localization. There are three different steps, and all three make use of magnetic gradient fields. This chapter starts with an overview of magnetic field gradients, and then we consider the first of the three steps required. This is slice selection. The principles are described and the interactive *Slice Selection* program is used to demonstrate how slice thickness and location can be selected.

5.1 Learning Outcomes

When you have worked through this chapter, you should:

- Be able to explain how gradient fields affect the main magnetic field in a magnetic resonance (MR) system
- Understand how the MR signal is localized to a selected slice through the body

5.2 Gradient Fields

Magnetic gradient fields are magnetic fields much smaller in size than the static uniform field. The small size of the gradient field is not the same everywhere within the main field. Instead, in the direction of the gradient, the field varies with location. It has a value of zero in the center, with a positive value at one extreme and a negative value at the other (Figure 5.1). The variation in the strength of a gradient field is expressed in units of mT m^{-1}. In Figure 5.1 the gradient is G mT m^{-1}. The gradient field strength has the value $-Gd$ at one end of the magnet and $+Gd$ at the other because in each case the displacement from the center of the magnet

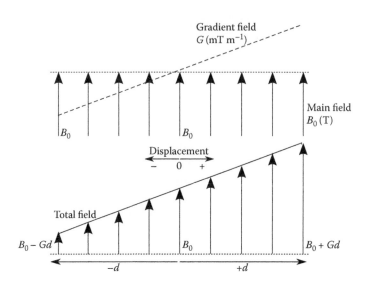

FIGURE 5.1

The strength of a gradient field is zero at the center and increases or decreases according to the displacement from the central location. The small gradient field strength is added to or subtracted from the main field strength, to give a resulting total field that varies with location. The arrows in this figure indicate the size of the main field, not its direction. The main field is always in the z direction.

is d m. The gradient field is superimposed on the main static field, which has the value B_0, and the result is a magnetic field with the value B_0 at the center, $B_0 - Gd$ at one end, and $B_0 + Gd$ at the other.

It is important to recognize that the direction of the magnetic field of the gradient is the same as the direction of the main field for the positive parts, and opposite the direction of the main field for the negative. The direction of the gradient itself, which is the direction in which it changes in value, not the direction of the magnetic field, can be aligned with any of the three axes, x, y, or z. The direction of the gradient in Figure 5.1 is the figure's horizontal direction.

The result of applying a gradient magnetic field is that the overall magnetic field strength is different at different spatial locations within the magnet. As a result, the Larmor frequency for the proton also differs, depending on the location in the magnet. This is a useful property for imaging, as it is possible to determine from where in the body a signal has been emitted from knowledge of the frequency. Figure 5.2 shows a gradient field applied to select a transverse slice and includes typical values of magnetic field strength.

Question 5.1

If a gradient runs in the x direction, in which direction should the magnetic field of the gradient run?

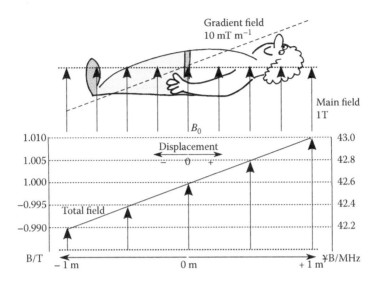

FIGURE 5.2
An example of a gradient field; in this case the gradient is being used for slice selection. The size of the gradient field strength is typical of those available. Values for the total magnetic field strength are indicated using both of the units, T and MHz, which are commonly quoted. The arrows in this figure indicate the size of the main field, not its direction. The main field is always in the z direction.

Answer

Whatever the direction of the gradient, the magnetic field of the gradient is always aligned with the main field, either parallel or antiparallel. It is conventional in magnetic resonance imaging (MRI) for the main field to run in the z direction.

Question 5.2

A gradient in the y direction has a field strength of 8 mT m^{-1} and its value at $y = 0$ m is 0 T.

(a) What is the value of the gradient field at $y = 0.5$ m?
(b) What is the value of the gradient field at $y = -0.25$ m?
(c) If the main field strength is 1 T, what is the total field at each of these locations?
(d) In what direction is the magnetic field of the gradient at each of these locations?

Answer

The strength of the gradient at $y = 0.5$ m is +4 mT, and at $y = -0.25$ m is –2 mT. These are obtained by multiplying together the y coordinate and the strength of the gradient. It is important to take care with positive and negative coordinate values, to ensure that the gradient field is correctly calculated with both positive and negative values, and is zero at the center (Figure 5.1). The total field is found by summing the main field and the gradient field, so the answers are 1.004 T and 0.998 T, respectively. For positive values of y the gradient field is in the +z direction, and for negative y the gradient field is in the –z direction.

Gradient strengths are usually expressed in units of mT m^{-1}. However, because magnetic field strength can equally well be expressed in units of frequency, with conversion via the Larmor equation, gradient strength is sometimes seen expressed as a number of MHz m^{-1} instead.

Worked Example

A brochure from a manufacturer states that the maximum gradient available in its system is 0.5 MHz m^{-1}. You require a system with a maximum gradient of at least 15 mT m^{-1}. Does the system meet your specification? What is the maximum gradient of that system expressed in mT m^{-1}? The gyromagnetic ratio $\gamma = 42.6$ MHz T^{-1}.

The relationship between field strength (B_0) and frequency (f_0) is expressed in the Larmor equation (Equation 3.4: $f_0 = \gamma B_0$), so this is the expression that will be applied to convert the gradient strength expressed in MHz m^{-1} to one expressed in mT m^{-1}. Dividing each side of Equation 3.4 by distance, and using the base SI units for frequency and magnetic field strength (i.e., Hz rather than MHz, T rather than mT), gives:

$$\text{Gradient field strength in Hz m}^{-1} = \gamma \times \text{gradient field strength in T m}^{-1}$$

The first option for determining whether the specification is met is to substitute the required value of 15 mT m^{-1} into the equation:

$$\text{Gradient field strength required} = 42.6 \times 10^6 \times 15 \times 10^{-3} \text{ Hz m}^{-1}$$
$$= 0.64 \text{ MHz m}^{-1}$$

So, the required gradient strength is greater than the maximum gradient offered (0.5 MHz m^{-1}), and so it is concluded that the system does not meet the specification.

To convert the maximum gradient strength given for the system into mT m^{-1}, rearrange the equation:

$$\text{Gradient field strength in T m}^{-1} = \text{gradient field strength in Hz m}^{-1}/\gamma$$

So,

$$\text{Gradient field strength provided} = 0.5 \times 10^6/42.6 \times 10^6 \text{ T m}^{-1}$$
$$= 12 \text{ mT m}^{-1}$$

Once again, it can be concluded that the system does not meet the specification, as 12 mT m^{-1} is less than the required 15 mT m$^{-1.}$

GRADIENT DEFINITIONS

In MRI, gradient fields are not applied all the time, but are switched on and off. Commonly encountered terms associated with switching are *rise time* and *slew rate*. The rise time is the time taken for the gradient strength to increase from zero to its final value, often a few hundred microseconds. The slew rate is found by dividing the final gradient strength by the rise time.

5.3 Gradient Fields for Slice Selection

5.3.1 Direction of Gradient for Slice Selection

To select a transverse slice across the body, a gradient is applied running from toe to head (Figure 5.3). The subject is lying supine, parallel to the direction of the main field, which runs in the z direction. This arrangement means that the precessional frequency is different at each location along the z axis, and so a particular frequency is associated with protons at each z location, that is, the protons in a transverse slice through the body. Slices can be selected in the other directions too—the direction of the gradient is always perpendicular to the direction of the required slice. Changing the strength of the gradient changes the thickness of the selected slice, and we shall return to this relationship shortly.

Question 5.3

In Figure 5.4, if the gradient is applied in the x direction, will a transverse, sagittal, or coronal slice be obtained?

Answer

A gradient in the x direction would run from side to side across the body in Figure 5.4. A sagittal slice would result because only protons with the same x location would share the same precessional frequency.

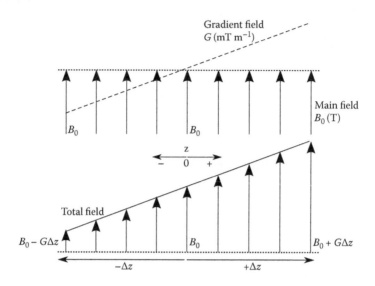

FIGURE 5.3

For selection of a transverse slice the gradient runs in the z direction. The resulting total field varies with location in the z direction, in this case from $B_0 - G\Delta z$ at the left to $B_0 + G\Delta z$ at the right. For a subject lying supine in the scanner, with his body aligned with the z axis, the z direction corresponds with the craniocaudal direction along his body. The arrows in this figure indicate the size of the main field and not its direction.

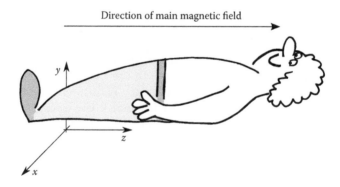

FIGURE 5.4

Diagram for Question 5.3.

Question 5.4

Three slices through the head are shown in Figure 5.5. The axis names are indicated on each image. Determine the direction of the gradient (i.e. x, y, or z) used to select the slice in each case.

Answer

The answers are (a) z, (b) x, and (c) y. The gradient direction is always perpendicular to the slice selected.

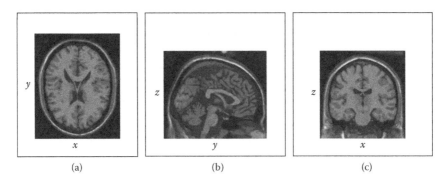

(a)   (b)   (c)

FIGURE 5.5

Labeled images for Question 5.4. (Image data from the BrainWeb Simulated Brain Database, www.bic.mcgill.ca/brainweb/; Collins, D. L., et al., Design and construction of a realistic digital brain phantom, *IEEE Trans. Med. Imag.*, 17, 463, 1998. With permission.)

5.3.2 Timing of Application of the Gradient Field for Slice Selection

The slice selection gradient is just one of three different gradients that are used for localization. The other two are discussed in Chapters 6 and 7. Each gradient is switched on at a different time in the imaging sequence (Figure 5.6). The slice selection gradient field is applied at the same time as the radio frequency (RF) pulse is applied to tip the net magnetization vector from its equilibrium position. Only those protons precessing at the same frequency as the RF pulse will be affected by the pulse, and these will be within a selected slice. The gradient is not needed once the RF pulse has been switched off, as slice selection has by then taken place. The only spins in the body that are affected by the RF pulse are those within the slice, and they are the only ones that will emit a signal.

The characteristics of both gradient and RF pulse affect the slice selection process. In the discussion so far we have implied that an infinitely thin slice could be selected, using an RF pulse containing a single frequency. In practice, the RF pulse itself contains a range of frequencies (this is called its *transmitted bandwidth*), and transmitted bandwidth has an effect on the thickness of the selected slice.

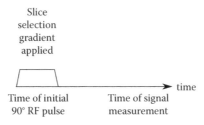

FIGURE 5.6

The slice selection gradient field is applied at the same time the RF pulse is applied.

5.4 RF Pulse for Slice Selection: Center Frequency and Transmitted Bandwidth

The RF pulse has two properties: its center frequency (f) and the transmitted bandwidth (TBW).

5.4.1 Effect of Center Frequency on Location of Slice

Changing the center frequency will change the location of the selected slice. This is illustrated in Figure 5.7. In the coordinate system used (Figure 5.3), displacement may be positive or negative and is measured from the isocenter of the magnet, where the gradient has a value of zero and does not affect the strength of the main field.

The center frequency for the RF pulse, f, is given by

$$f = f_0 + \Delta f \tag{5.1}$$

where f_0 is the frequency at the isocenter of the magnet and Δf is the offset of the center frequency from f_0,

$$\Delta f = G_{SS} \gamma d \tag{5.2a}$$

where G_{SS} is the slice selection gradient strength in Tm^{-1}, γ is the gyromagnetic ratio, and d is the displacement in meters from the isocenter of the magnet. Δf, f, and f_0 all have units of frequency (Hz). γ is usually given in MHz T^{-1}, so remember to account for the factor of 10^6 arising from the use of MHz instead of Hz when doing calculations.

Equation 5.2a may be rearranged:

$$d = \frac{\Delta f}{G_{SS} \gamma} \tag{5.2b}$$

From Equation 5.2b it can be seen that, for a particular slice selection gradient strength, the selected slice will be farther from the zero position for larger values of Δf. Similarly, for a fixed center frequency, d increases as the gradient is decreased.

There are limits on how far from the center of the magnet the slice can be shifted, mainly because the center frequency for the RF pulse has a maximum in a given MR system. However, slice selection also demands that the value of the main field is homogeneous, so that the total field is of the size expected. Similarly, the gradient must be linear right through the magnet if slices are to be selected correctly at all locations. Toward the edges of the magnet the main field is less homogeneous and gradient linearity may be worse, so slice selection far from the center is not reliable.

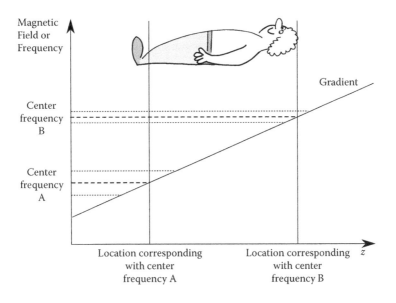

FIGURE 5.7

The choice of the center frequency of the RF pulse (dashed) determines the location of the selected slice. In this example, where the gradient increases from foot to head, lower center frequencies are associated with slices near the feet, and higher center frequencies with slices in the upper body. The dotted lines indicate the transmitted bandwidth.

Question 5.5

The slice select gradient is 10 mT m^{-1}. What is the offset from the frequency at the isocenter of the magnet for a slice centered 5 mm from the isocenter? The gyromagnetic ratio is $\gamma = 42.56$ MHz T^{-1}.

Answer

Substitute into Equation 5.2a, and the offset frequency is found to be 2.1 kHz.

5.4.2 Transmitted Bandwidth

The transmitted bandwidth is the range of frequencies in the RF pulse. A TBW of 1 kHz would correspond with, for example, a pulse centered at 42.6015 MHz and containing frequencies from 42.601 to 42.602 MHz. Note that an RF pulse of short duration has a wide TBW and a longer pulse has a narrower TBW.

Transmitted bandwidths can be expressed in units of frequency (Hz) or magnetic field strength (T). Conversion from one unit to the other is performed using the Larmor equation (Equation 3.4). For example, a TBW of 1 kHz may be expressed in units of magnetic field strength as 0.02 mT.

5.4.3 Effect of Gradient Strength and Transmitted Bandwidth on Slice Thickness

Changing the gradient strength or the transmitted bandwidth changes the thickness of the selected slice, which is the pixel size in the slice select direction.

The thickness of the selected slice in meters is given by

$$Pixel\ size_{SS} = \frac{TBW}{G_{SS}\gamma} \tag{5.3}$$

where TBW is the transmitted bandwidth in units of frequency, G_{SS} is the gradient strength in Tm^{-1}, and γ is the gyromagnetic ratio. γ is usually given in MHz T^{-1}, so remember to account for the factor of 10^6 arising from the use of MHz when doing calculations.

Equation 5.3 shows that, for a fixed gradient, the slice thickness increases if TBW is increased, and this relationship is illustrated in Figure 5.8. Similarly, for a fixed TBW, the slice thickness increases if the gradient strength is decreased (Figure 5.9).

There are limits on the thinness of a slice that can be acquired, because MR systems have both a maximum gradient strength (can be up to 50 mT m^{-1}) and a lower limit on the transmitted bandwidth.

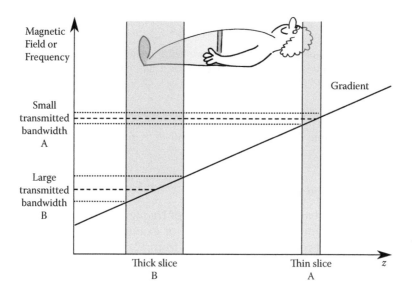

FIGURE 5.8
For a particular gradient, the slice thickness is determined by the TBW. The smaller the TBW, the smaller the range of frequencies present in the pulse, and the thinner the slice selected.

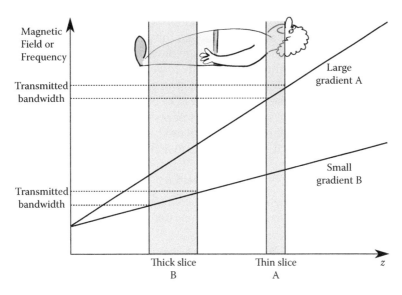

FIGURE 5.9
For a particular TBW, the slice thickness is determined by the size of the gradient. The larger the gradient, the thinner the slice selected.

Worked Example

If an MR system has a maximum gradient strength for slice selection of 30 mT m^{-1} and a minimum transmitted bandwidth of 0.8 kHz, what is the thinnest slice that can be selected? Assume that the gyromagnetic ratio γ = 42.6 MHz T^{-1}.

The thinnest slice will result when the TBW is as small as possible and G_{SS} is as large as possible. Substituting in Equation 5.3,

$$Pixel\ size_{SS\,min} = \frac{TBW_{min}}{G_{SS\,max}\,\gamma} \tag{5.4}$$

The minimum and maximum values required in Equation 5.4 were supplied in the question, so

$$Pixel\ size_{SS\,min} = 0.8 \times 10^3/(30 \times 10^{-3} \times 42.6 \times 10^6)\ m$$

$$= 6.26 \times 10^{-4}\ m\ (0.6\ mm)$$

5.4.4 Advantages and Disadvantages Associated with Thin Slices

We have seen that the slice thickness can be reduced using either a strong gradient or a narrow transmitted bandwidth. Strong gradients have a number of disadvantages, including the generation of eddy currents in parts of the magnet, which may lead to artifacts in the images. Also, mechanical vibrations and acoustic noise arise when the gradients are switched on or the direction is changed. The advantage of using a strong gradient rather than a narrow transmitted bandwidth to get a thin slice is that a short RF pulse can be used together with a strong gradient. A short RF pulse means, in turn, that the time to echo (TE) of the pulse sequence can be short. If a narrow TBW is used, this is necessarily associated with a long RF pulse, meaning that short TE times are not possible. If short TE times are not available, then it is not possible to achieve T1-weighted images.

Question 5.6

Which of the following is an alternative way of expressing the TBW for an RF pulse with frequencies from 60.0025 to 60.0045 MHz?

(a) 2 kHz
(b) 2 MHz
(c) 25 kHz
(d) 20 MHz
(e) 60.0035 MHz

Answer

The range of frequencies given is (60.0045 − 60.0025) = 0.002 MHz = 2 kHz. So, the only correct answer in the list is (a).

Question 5.7

Convert the following gradient strengths, which are expressed in terms of frequency, into mT m^{-1}. Assume that the gyromagnetic ratio $\gamma = 42.6$ MHz T^{-1}.

(a) 2 MHz m^{-1}
(b) 1 MHz m^{-1}
(c) 0.3 MHz m^{-1}

Answer

The answers are (a) 46.9 mT m^{-1}, (b) 23.5 mT m^{-1}, and (c) 7.0 mT m^{-1}.

Question 5.8

Calculate the slice thickness for each of the above gradients, if the TBW is 1.5 kHz.

Answer

The slice thicknesses found using Equation 5.3 are (a) 0.75 mm, (b) 1.5 mm, and (c) 5 mm. As expected, the slice thickness is greater for the smaller gradient strengths.

5.5 The Slice Selection Program

5.5.1 Slice Thickness

The Slice Selection program, which is on the CD, is a good way to become familiar with the effects of the transmitted bandwidth and gradient strength on the slice thickness. Start the program as explained in Chapter 1. The interface, shown in Figure 5.10, has four separate sets of controls. In each, you can choose if you want to select a transverse, sagittal, or coronal slice. As soon as the slice type has been selected, a diagram appears showing how the gradient strength and properties of the RF pulse affect the resulting slice. In the three areas at top left, top right, and bottom left, the gradient strength and transmitted bandwidth can be adjusted and the center frequency is fixed. The area at the bottom right is different from the other three. Here, the center frequency of the RF pulse can be changed instead of the transmitted bandwidth.

Set the program to select a transverse slice through the body by clicking on the word *Transverse* in the top-left display area. Remember that the gradient field direction is always perpendicular to the selected slice. This

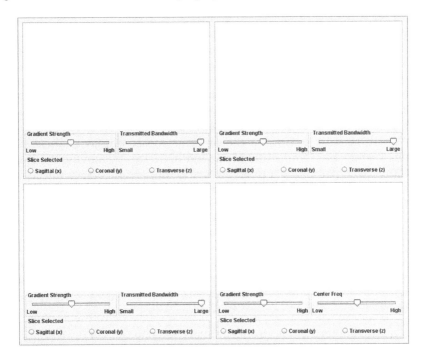

FIGURE 5.10
The interface for the Slice Selection program.

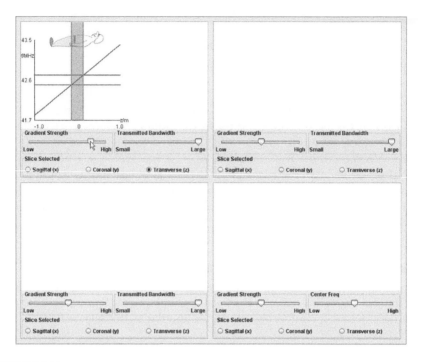

FIGURE 5.11
The Slice Selection program showing, in the upper-left quadrant, selection of a transverse slice. The gradient strength can be changed using the slider control.

is why the illustration of the body that is shown in the program is not a transverse slice, but is a toe-to-head view on which the gradient direction, in this case the z direction, can be seen.

We will start by setting up the program to use a high gradient field strength and a high transmitted bandwidth. Do this by moving the gradient strength slider for the top-left display to the far right of its range. Leave the transmitted bandwidth slider at the extreme right of its range. As you increase the gradient strength you should see the slice thickness, indicated by the shaded blue area, decrease (Figure 5.11). The diagram in the program is similar to those in the earlier parts of this chapter and those seen in other textbooks:

- The red line indicates how the size of the total magnetic field changes in value, in the z direction from the head to the toes of the patient in the scanner, because of the application of the gradient. Note that the field is stronger at the head end than at the feet.
- The vertical scale on the graph is given in terms of frequency. Viewing this scale helps us to see how the precessional frequency varies with location in the z direction. The values can be expressed

in units of magnetic field strength instead (see "Applying the Larmor Equation" box).

- The green lines show the transmitted bandwidth, which is the range of frequencies present in the RF pulse that is used to disturb the spins from their equilibrium precession. Only the spins in the body that are precessing with the particular frequencies in the range of the transmitted bandwidth will be affected by the RF pulse.

- The blue lines show the range of z positions where spins have the precessional frequencies that are present in the transmitted bandwidth. Only spins located between the blue lines will be affected by the RF pulse; in other words, we have selected a slice with the thickness of the gap between the blue lines. Note that projections of the blue lines across the body image cut through the body shown to indicate the location of the transverse slice.

APPLYING THE LARMOR EQUATION

The magnetic field axis in the Slice Selection program uses units of frequency to indicate the precessional frequency of protons at different z locations in the field. The values can be expressed in units of magnetic field strength instead by using the Larmor equation, which states that there is a linear relationship between magnetic field strength and frequency of precession. To convert from frequency to field strength using the Larmor equation, use Equation 3.4, where the gyromagnetic ratio has a value of 42.6 MHz T^{-1} for clinical MRI using the proton.

Question 5.9

This question will ensure that you know what each of the features shown on the plot in the Slice Selection program represents. Delete the wrong answers:

The total magnetic field strength is shown by the red/green/blue lines.
The transmitted bandwidth is shown by the red/green/blue lines.
The selected slice is indicated by the red/green/blue lines.

Answer

You should have answered that the total magnetic field strength is shown in red, the transmitted bandwidth by the green lines, and the selected slice by the blue lines.

Question 5.10

In the example shown in Figure 5.11, the range of values shown on the frequency axis of the plot is from 41.7 MHz to 43.5 MHz. With what magnetic field strengths do these two values correspond?

Answer

The frequency axis shows the precessional frequency for the proton, so the corresponding magnetic field strength in tesla is found using the Larmor equation. As the frequency is expressed in MHz, the correct formulation to use is Equation 3.4:

$$f_0 = \gamma B_0$$

where f_0 is the precessional frequency in MHz, γ is the gyromagnetic ratio for the hydrogen nucleus (42.6 MHzT^{-1}), and B_0 is the magnetic field strength in T.

$$\text{If } f_0 = 41.7 \text{ MHz}, B_0 = \frac{f_0}{\gamma} = 41.7/42.6 = 0.979 \text{ T}$$

$$\text{If } f_0 = 43.5 \text{ MHz}, B_0 = \frac{f_0}{\gamma} = 43.5/42.6 = 1.021 \text{ T}$$

It can be seen from these values that typical gradients need only be relatively small; changing the field by a few mT is enough to change the precessional frequency by more than 1 MHz.

Now, we will generate more plots to compare with the first one, all involving selection of a transverse slice:

In the upper-left quadrant:
- Leave the *Transverse* slice selected.
- Ensure that the gradient strength slider is at the extreme right of its range.
- Leave the transmitted bandwidth slider at the extreme right of its range.

In the upper-right quadrant:
- Click on the word *Transverse*.
- Move the gradient strength slider to the middle of its range.
- Leave the transmitted bandwidth slider at the extreme right of its range.

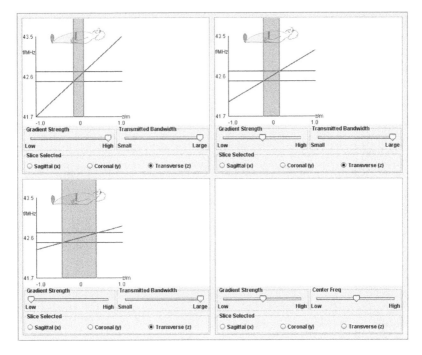

FIGURE 5.12

The Slice Selection program showing selection of a transverse slice in three quadrants, in each case using the same transmitted bandwidth but a different gradient strength.

In the lower-left quadrant:

- Click on the word *Transverse*.
- Move the gradient strength slider to the extreme left of its range.
- Leave the transmitted bandwidth slider at the extreme right of its range.

Your display should appear as shown in Figure 5.12. These selections mean that the only thing that is different in the three plots is the magnetic field strength. The three diagrams show that decreasing the gradient increases the width of the selected slice.

Slice selection gradients can also be used to select coronal and sagittal slices. To get a display that shows the effect of gradient strength on slice thickness for sagittal slices, click on the word *Sagittal* in the slice selected area of all three quadrants. Do not make any changes to the gradient strengths and transmitted bandwidths selected. Your display should appear as shown in Figure 5.13. A sagittal slice is one that divides the body from side to side, so the illustrated body view has changed to one that includes the two sides of the body, viewed from the feet. The gradient

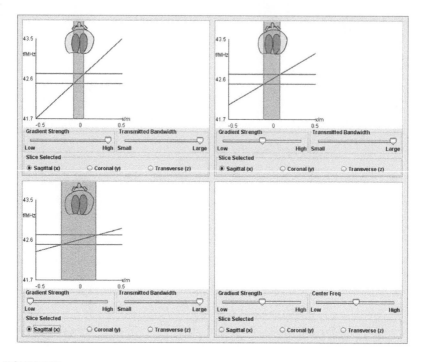

FIGURE 5.13
The Slice Selection program showing selection of a sagittal slice in three quadrants, in each case using the same transmitted bandwidth but a different gradient strength.

for a sagittal slice runs in the x direction (notice that the axis label has changed), and across the body from side to side. This is in contrast to the gradient for the transverse slice, which ran from head to toe.

Question 5.11

Make the correct selections in the slice selection program to obtain a display that matches the one shown in Figure 5.14.

Answer

The display shown in Figure 5.14 was achieved by clicking on the word *Coronal* in the slice selected area of all three quadrants without making any changes to the gradient strengths and transmitted bandwidths selected. For the coronal slice notice that the axes of the graph and illustrated view of the body have been rearranged to emphasize that the gradient field runs in the y direction, from front to back across the body.

In the first exercises with the slice selection program, we kept the transmitted bandwidth fixed and changed the gradient strength. Now, in each of the three quadrants, request a transverse slice, select the same gradient

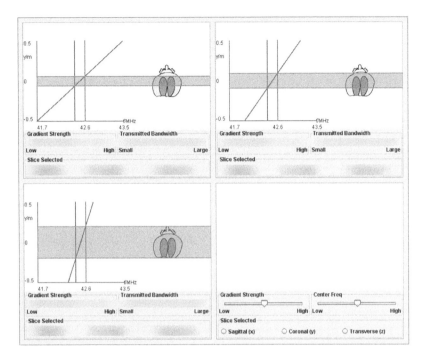

FIGURE 5.14
Image for Question 5.11.

strength for all three, but select a different (small, medium, and large) transmitted bandwidth for each of the quadrants. The display should look similar to the one shown in Figure 5.15. It may differ slightly if you chose a different gradient strength, but as long as the same gradient strength is used for each plot, it does not matter which one it is. We can see how, for a fixed gradient strength, the selected slice thickness is changed by adjusting the transmitted bandwidth. The greater the transmitted bandwidth (the range of frequencies in the pulse), the thicker is the selected slice.

Question 5.12

Think about which combination of gradient strength and transmitted bandwidth you would expect to give the thinnest slice and the thickest slice, and try the combination in the program.

Answer

From Equation 5.3, the thinnest slice arises from a high gradient strength and a small transmitted bandwidth, and the thickest slice from a low gradient strength and a large transmitted bandwidth. The results of selecting these combinations in the slice selection program are shown in Figure 5.16.

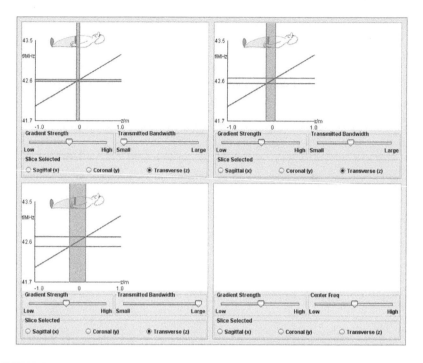

FIGURE 5.15
The Slice Selection program showing selection of a transverse slice in three quadrants, in each case using the same gradient strength but a different transmitted bandwidth.

5.5.2 Slice Position

Finally, use the bottom-right quadrant to see how the choice of the center frequency of the transmitted RF pulse affects the position of the selected slice. In this quadrant the transmitted bandwidth has been fixed to the small value that leads to the thinnest slice.

In the lower-right quadrant:

- Click on the word *Transverse*.
- Move the gradient strength slider to the extreme right of its range.
- Move the center frequency slider to the extreme left of its range.

Move the center frequency slider slowly from left to right. You will see that as the center frequency increases, the selected slice moves between the feet and the head of the subject (Figure 5.17a,b,c).

The size of the difference between the center frequencies used to select adjacent slices determines if the slices will overlap, be contiguous, or have a gap between them.

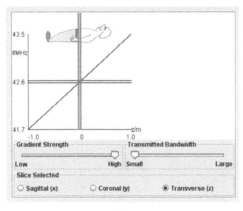

(a)

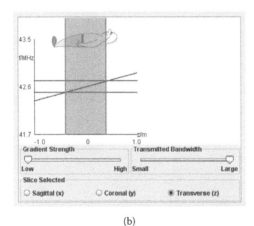

(b)

FIGURE 5.16
Feedback for Question 5.12. The Slice Selection program showing selection of (a) a thin transverse slice using a high gradient strength and a small transmitted bandwidth, and (b) a thick transverse slice using a low gradient strength and large transmitted bandwidth.

Question 5.13

Use the lower-right quadrant of the slice selection program to deduce how far apart the center frequencies used to select contiguous slices must be. Your answer will be in terms of other parameters, such as gradient strength or transmitted bandwidth; a numerical answer is not expected.

Answer

Contiguous slices result when the successive transmitted bandwidths are contiguous. This corresponds with center frequencies spaced by the value of the TBW.

TO CLOSE THE PROGRAM

Click on the cross at the top right of the window, and then confirm that you want to end the program.

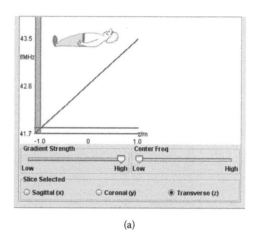

(a)

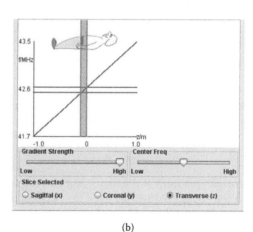

(b)

FIGURE 5.17

The Slice Selection program showing selection of a transverse slice in different locations by changing the center frequency. (a) Low center frequency, (b) medium center frequency, (c) high center frequency.

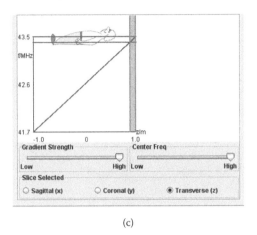

(c)

FIGURE 5.17   (*continued*)

5.6 Acquiring Several Slices

A slice profile plots the strength of the signal through a slice in a direction at right angles to the plane of the slice. It is assumed that the imaged material is homogeneous. For a transverse slice the profile is plotted through the slice in the z direction. Ideally, the slice profile would be rectangular, as shown in Figure 5.18, indicating that exactly the same signal is acquired from all points within the slice, and no signal is acquired from any point outside the slice. The profile has sharp edges and is uniform from edge to edge. In reality, the slice profile is likely to have sloping

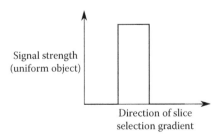

FIGURE 5.18

A rectangular slice profile. The profile plots the signal strength through the slice, in a direction perpendicular to the slice plane. In the ideal case of the rectangular slice profile shown here, the same signal is acquired from all points within the slice, and no signal is acquired from outside the slice.

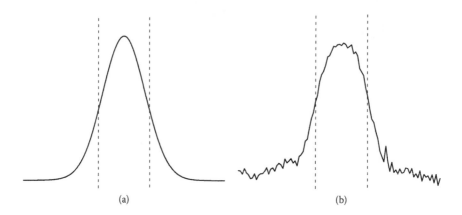

FIGURE 5.19
Examples of slice profiles, both with a width of 5 mm. In each case, the position of a rectangular slice of thickness 5 mm and centered in the same place is indicated by the dashed lines. (a) Gaussian profile, (b) a realistic example of a 5-mm-thick transverse slice profile.

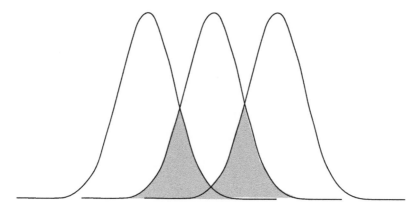

FIGURE 5.20
Cross talk, indicated by shading, arises when contiguous slices with nonrectangular profiles are acquired.

edges and be nonuniform across the slice. A Gaussian slice profile is shown in Figure 5.19a, and an example of a realistic slice profile appears in Figure 5.19b. The lack of sharp edges means that if contiguous slices are acquired, there will be overlap (Figure 5.20) at the edges of the real slice profiles. This effect is called *cross talk*. The tissue in the overlap regions will be excited by the acquisitions for both slices, the effective TR interval is smaller than intended, and the tissue does not have time to fully relax. As a result, there is reduced signal and increased signal-to-noise ratio (SNR) in the overlap regions. To avoid cross talk between slices, slices are acquired farther apart (gapped) by increasing the separation of center frequencies. Alternatively, the acquisitions can be interleaved in time, so that adjacent slices are not acquired together.

5.7 Additional Self-Assessment Questions

Question 5.14

(a) Sketch the diagram and graph in Figure 5.21. Assume that the system has a horizontal magnetic field that runs in the direction of the bore of the magnet. On your sketch of the diagram indicate the direction of the magnetic field with an arrow.

(b) If no gradient fields are applied, complete the graph to show how the value of B_0 varies with position along the bore of the magnet.

(c) Add a second arrow to the diagram to indicate the direction in which the gradient must run to select a transverse slice. Label each end of the arrow to show where the resulting field is low and where it is high.

(d) Add to the graph to show how the strength of the magnetic field varies with position along the bore of the magnet when the gradient is applied.

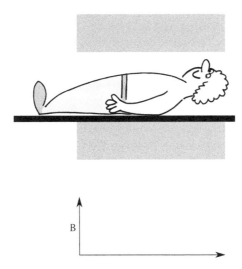

FIGURE 5.21
Base diagram to be used for Question 5.14.

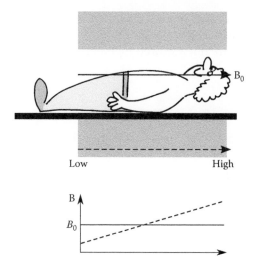

FIGURE 5.22
Feedback on Question 5.14.

Answer

Figure 5.22 shows a completed diagram. (a) The arrow indicating the direction of the magnetic field should be horizontal and run from head to toe (or toe to head) of the subject. (b) The value of B_0 when no gradients are applied is constant. This is shown by the solid line in the graph at the bottom of the figure. (c) To select a transverse slice, the gradient must vary from head to toe or from toe to head. Any other direction of variation would select a slice from a different plane. This is shown by the dashed arrow. In this example the field is labeled low at the foot end and high at the head end. It would also be correct for the labels to be the other way round. (d) The dashed line in the sketch graph shows the field strength slightly smaller than B_0 at the toes and slightly higher than B_0 at the head. If you chose to make the field low at the head end, then in your answer the line in the sketch graph should slope in the opposite direction. The direction of the field itself is the same as that of the main static field B_0. The gradient has added or subtracted only a fraction of the field strength and does not change the direction of the main field. The size of the field varies in the direction required.

Question 5.15

It is desired to select a slice using a gradient field as indicated in Figure 5.23. Will the proposed gradient result in a transverse, sagittal, or coronal slice?

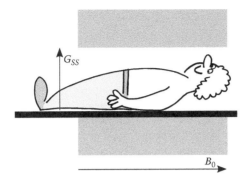

FIGURE 5.23
Diagram for Question 5.15.

Answer

The gradient changes the value of the magnetic field in the posterior-anterior direction, and so will result in a coronal slice.

Question 5.16

In the diagram in Figure 5.24, which of the two alternatives, (a) or (b), will lead to selection of a thinner slice, assuming a fixed transmitted RF bandwidth?

Answer

In Figure 5.24, option (b) has the steeper gradient. For a fixed bandwidth in the transmitted RF pulse, the range of frequencies will be covered within a shorter length of body and lead to a thinner slice.

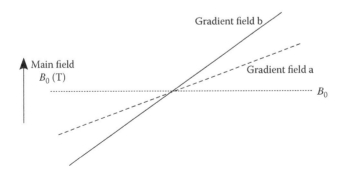

FIGURE 5.24
Diagram for Question 5.16.

Question 5.17

Select the correct option from the following alternatives. To perform slice selection, the slice select gradient is turned on:

(a) At the same time as the transmission of the RF pulse.
(b) Immediately after transmission of the RF pulse.
(c) At the time the signal is measured.
(d) For the whole period between transmission of the RF pulse and measurement of the signal.

Answer

The correct answer is (a). The slice select gradient is only required at the time of the transmission of the RF pulse. This means that only those spins in the slice of interest are affected by the RF pulse and will ultimately generate a signal. In practical use, compensatory gradients are sometimes applied at other times during the pulse sequence to correct for unwanted dephasing caused by the gradients used for spatial localization.

Question 5.18

Explain how a transverse slice in a different location can be selected that has the same slice thickness as the original slice.

Answer

The same gradient strength is used and the RF pulse must have the same bandwidth to keep the slice thickness the same. To acquire a different slice, the RF pulse must have a different center frequency.

Question 5.19

Which of the following statements are true? Thicker slices can be achieved by:

(a) Increasing the transmitted bandwidth of the RF pulse.
(b) Increasing the center frequency of the RF pulse.
(c) Decreasing the slice select gradient strength.
(d) Lengthening the RF pulse.
(e) Increasing the slice select gradient strength.

Answer

The two ways to change the slice thickness are by changing the slice select gradient strength and by adjusting the transmitted bandwidth of the RF pulse. The correct answers are (a) and (c). Lengthening the RF pulse corresponds with decreasing the transmitted bandwidth, which would reduce the slice thickness.

5.8 Chapter Summary

This chapter began with a general explanation about the use of gradient fields in MRI, and how they are used to make small changes to the strength of the main field so that the field strength changes in the chosen direction. The use of a gradient field to select a slice of tissue for MRI was described next. Slice selection is the first of the three steps that are used to localize the signal and eventually give an MR image. The effect of the gradient field strength and transmitted bandwidth on the slice thickness were discussed, together with the role of the center frequency in determining the slice location. These relationships were further demonstrated using the slice selection program. In the next chapter the next step in the localization procedure will be covered: frequency encoding.

6

Frequency Encoding

In the previous chapter the process of slice selection was described. Slice selection means that it is known from what slice of tissue a magnetic resonance (MR) signal is originating. The frequency encoding procedure, which is the subject of this chapter, allows the signal to be further localized to columns within the slice. The principles of frequency encoding are described, and the interactive *Frequency Encoding Demonstrator* program is used to illustrate these principles using graphics.

6.1 Learning Outcomes

When you have worked through this chapter, you should:

- Understand how the MR signal is localized to a particular column (or row) in the selected slice through the body
- Be able to describe how the values of the parameters that are associated with frequency encoding affect the field of view

6.2 Principles of Frequency Encoding

In slice selection, covered in the previous chapter, the location of the slice was determined by the value of the center frequency of the radio frequency (RF) pulse, and the magnetic field gradient was used to limit the volume in which tissue would be excited and so later emit a signal. In frequency encoding, which is the topic of this chapter, the magnetic field gradient is used in an analogous way. This time the gradient leads to a variation in the resonance frequency across the slice, so that within the previously excited slice the signal emitted from each different location has a different frequency.

If the slice selection gradient was applied in the z direction, then the frequency encoding gradient may be in either the x or y direction. We shall use the x direction in this chapter. Frequency encoding will allow

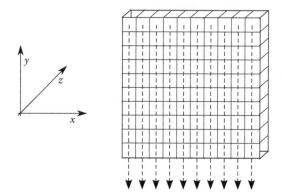

FIGURE 6.1
Frequency encoding in the *x* direction localizes the signal to a column in the previously selected slice. In this example a slice has been selected using a gradient in the *z* direction. The dashed arrows indicate that each column in the slice is associated with a location in the *x* direction.

determination of the position of the signal in the *x* direction, corresponding with columns in the image (Figure 6.1). We are not yet able to isolate the location in both *x* and *y* directions—this needs the third step of spatial localization, phase encoding, which is covered in the next chapter. The gradient for frequency encoding is applied at the time of signal measurement, which differs from slice selection, where the gradient field is applied at the time of the RF pulse.

6.3 Gradient Fields for Frequency Encoding

6.3.1 Direction of Gradient for Frequency Encoding

The magnetic field gradient for frequency encoding works in exactly the same way as the gradient used for slice selection, but it is applied in a different direction and at a different time in the pulse sequence. To apply frequency encoding from side to side across the body in a transverse slice, the slice selection gradient is applied running from toe to head, and the frequency encoding gradient from left to right. The precessional frequency is different at each location along the *x* axis, and so a particular frequency is associated with protons at each *x* location (Figure 6.2). The slice might be selected using a gradient in a direction other than *z*, but the frequency encoding gradient is always applied in a direction perpendicular to the slice selection gradient. Changing the strength of the frequency encoding gradient changes the extent of tissue imaged in the direction of the frequency encoding gradient—the *field of view* (FOV).

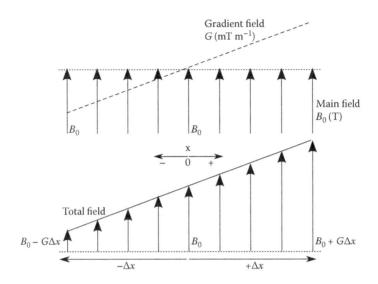

FIGURE 6.2

For frequency encoding of a transverse slice the frequency encoding gradient may run in the x direction or the y direction. An x gradient is shown here. The resulting total field varies with location in the x direction, in this case from $B_0 - G\Delta x$ at the left to $B_0 + G\Delta x$ at the right. For a subject lying supine in the scanner, with his body aligned with the z axis, the x direction corresponds with the lateral direction from side to side across his body. The arrows in this figure indicate the size of the main field and not its direction.

Question 6.1

A slice selection gradient has been used to select a coronal slice.

(a) Does this slice selection gradient run in the head-to-toe, front-to-back, or side-to-side direction through the body?
(b) In which of the three directions through the body is it possible to apply frequency encoding for this coronal slice?

Answer

The selected slice is a coronal slice, which separates the front and back of the body. So the slice selection gradient must run in the front-to-back direction. Frequency encoding may be applied in either of the two directions perpendicular to this, and so could run either from head to toe or from side to side.

6.3.2 The Composite Signal from Frequency Encoding

The frequency encoding gradient results in the emission of signals from the slice, each of which has one frequency from a range of frequencies. The signal detected is a composite signal representing the sum of all these individual signals. Each signal has a frequency associated with

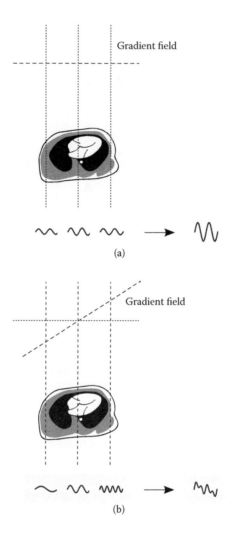

FIGURE 6.3
In frequency encoding, signals with different frequencies are emitted from different locations. The amplitude of the signal at each location depends on the number of protons emitting a signal (i.e., the amount of tissue present) at the location. A composite signal is detected
that is the sum of all the different signals. (a) With no applied gradient, the emitted signal
has the same frequency across the slice. (b) When a frequency encoding gradient is applied,
signals with different frequencies and amplitudes are emitted from different locations.

its x location, and has an amplitude that depends on the number of protons at the relevant x location. Two examples are shown in Figure 6.3. In
Figure 6.3a there is no gradient applied, and the precessional frequency
is the same across the slice. The measured signal contains only this frequency. In Figure 6.3b, a frequency encoding gradient has been applied,

and protons at different x locations are precessing with different frequencies. The measured signal is a more complicated shape, as it represents the sum of individual signals with different amplitudes and frequencies. This summing of the signals may suggest that we have lost the spatial localization information, but the Fourier transform is able to divide a composite signal like this into its component frequencies, so the information about the strength of the signal for each frequency is easily recovered.

6.3.3 Timing of Application of the Gradient Field for Frequency Encoding

The frequency encoding gradient is applied at the time the signal is measured (Figure 6.4). This differs from the gradient for slice selection, which was applied at the time of the RF pulse, as outlined in Chapter 5. The properties of the transmitted RF pulse are not important for frequency encoding, and instead the characteristics of the receiver come into play. The *receiver bandwidth* (RBW) defines the range of frequencies that can be detected in the measured signal, and will differ from system to system.

6.3.4 Receiver Bandwidth

The receiver bandwidth is usually given in kHz. The RBW gives the range of frequencies present across the whole of the field of view in the frequency encoding direction. Together with the gradient strength, the RBW determines the FOV; we shall return to this relationship later. A typical RBW is 80 kHz, meaning that field strengths will run from 40 kHz less than γB_0 at one extreme of the FOV to 40 kHz above γB_0 at the other side of the FOV (Figure 6.5). Just as it is acceptable to express magnetic field strengths and gradients in units of frequency as well as field strength, bandwidth may alternatively be expressed in tesla by applying the Larmor equation (Equations 3.1 and 3.4). For example, the RBW of 80 kHz corresponds with field strengths extending across $(80,000/\gamma)$ T $= 1.88$ mT.

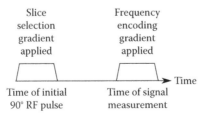

FIGURE 6.4
The frequency encoding gradient field is applied at the same time as the signal is measured.

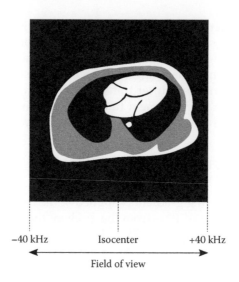

−40 kHz   Isocenter   +40 kHz

Field of view

FIGURE 6.5

The RBW indicates the range of frequencies present across the whole field of view in the frequency encoding direction. In the center of the FOV, at the isocenter, the precessional frequency is γB_0. In this example an RBW of 80 kHz reduces the precessional frequency by 40 kHz at one side of the FOV and increases the precessional frequency by 40 kHz at the other side of the FOV. The range of frequencies is from $\gamma B_0 - 40$ kHz to $\gamma B_0 + 40$ kHz.

6.4 The Frequency Encoding Demonstrator

The Frequency Encoding Demonstrator program, which is on the CD, can help you to become familiar with the ideas behind frequency encoding, which were summarized at the start of this chapter. Start the program as explained in Chapter 1. The appearance of the interactive demonstrator is shown in Figure 6.6. A section through the body is shown, with the x axis running from side to side, and the y axis in the anterior-posterior direction. The z axis is at right angles to both x and y and runs into the screen. This is a conventional arrangement of axes for a horizontal bore magnet, though sometimes the x and y directions might be swapped. The position where $(x, y, z) = (0, 0, 0)$ is the *isocenter* of the magnet.

Five differently colored vertical lines are shown that represent different values of x. There is a strong static magnetic field applied, which is indicated by the slider at the left, on which the field is set at 1.5 T. Associated with each location is a waveform in a matching color representing the frequency of precession, and hence of the emitted signal, at that x location. The amplitude of each signal is slightly different, indicating the differing amount of tissue that is emitting a signal at each x location. The greatest amplitude is associated with the orange location, which is associated

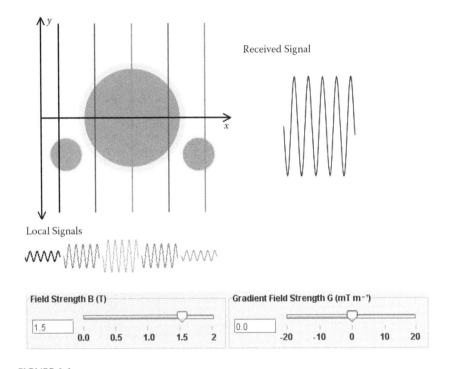

FIGURE 6.6
Screen shot showing the appearance of Frequency Encoding Demonstrator when the program is first run. In the demonstrator the vertical lines that indicate location are displayed in color (from left to right): black, red, orange, green, and blue.

with the largest body thickness. The interactive demonstrator allows you to visualize the effect on these frequencies of changing the strength of the gradient field used for frequency encoding in the x direction. The slider at the right can be used to change the strength of the gradient field. When the demonstrator starts up, there is no applied gradient, so the precessional frequency is the same at all the different x locations. The total signal received from the slice is represented by the black waveform on the right side of the demonstrator. At present, because there is no frequency encoding gradient applied, the signal is a simple sine wave containing just the one frequency. Note that the receiver bandwidth in the demonstrator is fixed.

Adjust the strength of the gradient by typing 10 into the box beside the right-hand slider, and use the return or enter key to make the change take effect (Figure 6.7). The signal frequencies illustrated at the five locations are now different from each other. The frequency at the center is unchanged, since the gradient field has a value of 0 mT at that location, but the frequency of precession in the blue location at the right is higher than in the black location at the left of the display, because the magnetic field is stronger at the right of the display.

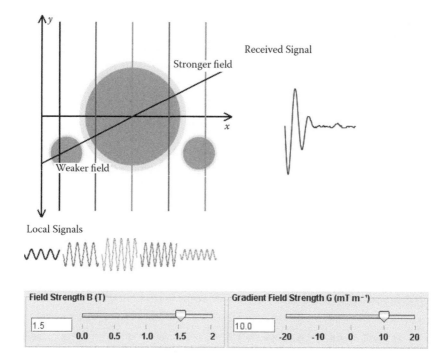

FIGURE 6.7
The Frequency Encoding Demonstrator with a gradient of 10 mT m⁻¹ selected.

Note how the introduction of the gradient has changed the composite signal (shown at the right of the window) from one with a single frequency to a more complicated waveform. The composite signal shown in the demonstrator is considerably simplified compared with the real-life case, as it shows the signal from only the five different x locations. In reality, signals will be detected from as many x locations as there are columns in the image, perhaps 256 or 512. Use the slider to change the value of the gradient strength and watch the resulting changes to the precessional frequencies and the composite signal. When using the slider, you will notice that negative gradient values are possible too, which simply result in a stronger field on the other side of the body than is the case for a positive gradient (Figure 6.8).

Now set the gradient to 0 mT m⁻¹ again. To demonstrate the effect of the main field on the precessional frequency, move the slider to select a different value of B_0. Then adjust the gradient with the other slider. You will see that for all nonzero values of the main field, adding a gradient always results in a precessional frequency that is related to the x location.

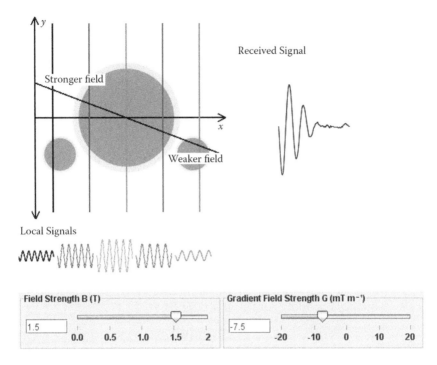

FIGURE 6.8

If a negative value is selected for the gradient, the stronger field occurs on the left side of the diagram. The stronger field is on the right for a positive gradient.

SIGNAL SIZE

When the main field B_0 is zero, the demonstrator shows no signal, to emphasize the need for a large static magnetic field to cause precession. The emphasis in the demonstrator is on differences in the signal at different locations, so it does not also include the increase in the strength of the signal that is expected with increase in the strength of the main magnetic field.

The composite signal is measured, and if a one-dimensional Fourier transform is applied, then the signals associated with particular frequencies, and with particular x locations, can be extracted. An image cannot be calculated yet, because although we know the x locations, there is no information regarding the y location of the measured signals. Further encoding is required, and this is phase encoding, which is covered in Chapter 7.

TO CLOSE THE PROGRAM

Click on the cross at the top right of the window, and then confirm that you want to end the program.

6.5 Effect of Gradient Strength and Receiver Bandwidth on Field of View (FOV)

6.5.1 Field of View

We have already introduced the field of view as a quantitative term used to describe the extent of tissue imaged in the direction of the frequency encoding gradient. FOV is measured in meters. The name *field of view* can also be used descriptively, to refer to the area or volume that is imaged, but it is the quantitative usage that is applicable in this section. The FOV in the frequency encoding direction is defined as

$$FOV_{FE} = \frac{RBW}{G_{FE}\gamma} \tag{6.1}$$

where RBW is the receiver bandwidth expressed in frequency units, G_{FE} is the frequency encoding gradient strength in T m^{-1}, and γ is the gyromagnetic ratio. The value for γ is usually given in MHz T^{-1}, so remember to include the factor of 10^6 arising from the use of MHz when doing calculations. From Equation 6.1 it can be seen that a small FOV is associated with either a large gradient (Figure 6.9a) or small receiver bandwidth (Figure 6.9b).

6.5.2 Pixel Size

If the number of pixels across the FOV is fixed, then a smaller FOV will correspond with smaller pixels, and thus with high spatial resolution. The pixel size can be found from knowledge of the FOV and the number of pixels in the frequency encoding direction:

$$Pixel\ size_{FE} = \frac{FOV_{FE}}{N_{FE}} \tag{6.2}$$

Substituting for FOV from Equation 6.1 in Equation 6.2 gives:

$$Pixel\ size_{FE} = \frac{RBW}{\gamma G_{FE} N_{FE}} \tag{6.3}$$

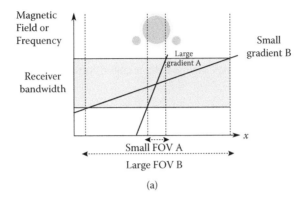

(a)

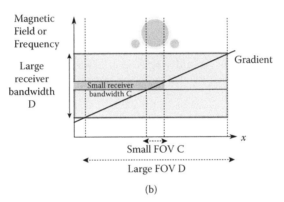

(b)

FIGURE 6.9
The effect of gradient field strength and receiver bandwidth on the FOV in frequency encoding. Previous selection of a transverse slice is indicated by the stylized cross section of the body. (a) For a fixed receiver bandwidth, a large gradient gives a small FOV (A). (b) For a fixed gradient strength, a small receiver bandwidth gives a small FOV (C).

From this expression it can be seen that the pixel size depends on the receiver bandwidth, the gradient strength, and the number of pixels required. For a fixed number of pixels, the smallest pixels are obtained by using a small receiver bandwidth and a large frequency encoding gradient.

Worked Example

If an MR system has a maximum gradient strength for frequency encoding of 10 mT m^{-1} and a minimum receiver bandwidth of 50 kHz, estimate the smallest FOV that is possible in the frequency encoding direction. The gyromagnetic ratio $\gamma = 42.6$ MHz T^{-1}.

The minimum FOV will result when RBW is as small as possible and G_{FE} is as large as possible. Substituting in Equation 6.1,

$$FOV_{FEmin} = \frac{RBW_{min}}{G_{FEmin} \gamma}$$
(6.4)

The minimum and maximum values required in this expression were supplied above, so

$$FOV_{FEmin} = 50 \times 10^3/(42.6 \times 10^6 \times 10 \times 10^{-3}) \text{ m}$$

$$= 0.12 \text{ m (or 12 cm)}$$

Question 6.2

Consider the system described in the worked example. Would the minimum FOV increase or decrease if stronger gradients were installed?

Answer

The minimum FOV would decrease if stronger gradients were installed; this can be deduced from the Equation 6.4 derived in the worked example.

6.5.3 Advantages and Disadvantages of a Small FOV

We have seen that the advantage of a small FOV is increased spatial resolution. However, when the FOV is smaller than the object being imaged, an imaging artifact occurs. The artifact is known as the wraparound or aliasing artifact. The principles of its generation are illustrated in Figure 6.10, and an example is shown in Figure 6.11.

6.6 Additional Self-Assessment Questions

Question 6.3

What is the FOV in the frequency encoding direction if $G_{FE} = 0.5$ mT m^{-1} and RBW = 20 kHz?

Answer

By substituting in Equation 6.1, FOV = 0.94 m. If you got the wrong answer, check that the correct powers of 10 were used.

Question 6.4

What is the pixel size in the x (frequency encoding) direction for a 256×256 image if $G_x = 0.5$ mT m^{-1} and RBW = 20 kHz?

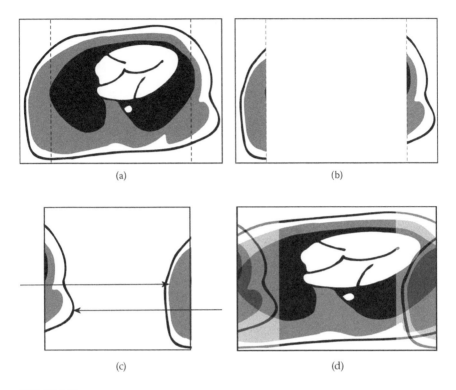

FIGURE 6.10
The wraparound imaging artifact occurs when the field of view is smaller than the object being imaged. Parts of the object that fall outside the field of view are reproduced inside the field of view and are on the opposite side of the image from their physical location. (a) The edges of the FOV are indicated with dashed lines. (b) These parts of the body are outside the FOV. (c) Aliasing causes the structures outside the FOV to be imaged inside the FOV and on the opposite side. (d) The resulting image includes both the structure from within the FOV and the aliased data.

Answer

The gradient strength and receiver bandwidth are the same as in the preceding question, so the FOV is again 0.94 m. By substituting in Equation 6.2, the pixel size obtained is 3.7 mm.

Question 6.5

What is the pixel size in the x (frequency encoding) direction for a 128×128 image if $G_x = 2$ mT m^{-1} and RBW $= 0.5$ mT?

Answer

Note that the receiver bandwidth was given in units of magnetic field strength rather than units of frequency. To convert from magnetic field strength to frequency multiply by γ (see Section 6.3.4). The

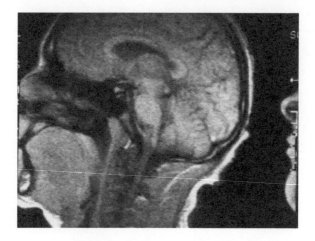

FIGURE 6.11
In this example of the wraparound artifact, the aliasing is seen at the right of the image. The nose is imaged behind the head rather than in the front. The artifact arose because the nose was outside the field of view at the time of imaging. (Image courtesy of John Ridgway.)

answers, using Equations 6.1 and 6.2, are $FOV_{FEx} = 0.25$ m and pixel size = 1.95 mm.

Question 6.6

Figure 6.12a shows a screen shot of the frequency encoding demonstrator. The waveforms that usually appear under the five vertical location lines have been removed from the image and are shown in Figures 6.12b to f, where they are all drawn in black and appear in a different order than they did in the demonstrator. What is the order of the waveforms, from left to right, to correspond with the screen shot?

Answer

The correct order is e, d, b, f, c.

Question 6.7

An MR system has a maximum gradient strength of 10 mT m^{-1} and a minimum receiver bandwidth of 80 kHz. Estimate the smallest FOV that is possible.

Answer

From Equation 6.4, the minimum FOV achievable in the frequency encoding direction is 18.8 cm.

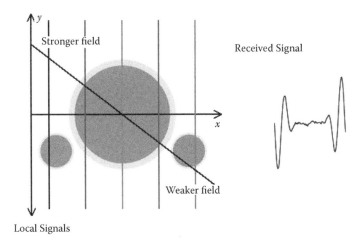

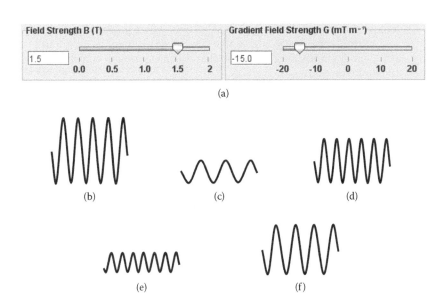

FIGURE 6.12
Images for Question 6.6.

Question 6.8

Select the one true statement from the following options. The frequency encoding gradient is applied:

(a) At the same time as the slice select gradient.
(b) At the same time as the signal is measured.

   (c) After the slice selection gradient, but before the signal measurement takes place.

   (d) During the 90° RF pulse that causes saturation.

   (e) Immediately after the 90° RF pulse that causes saturation.

Answer

The true statement is (b).

Question 6.9

An MR system has a transmitted bandwidth of 0.5 mT and a receiver bandwidth of 18 kHz. The slice select gradient is set to 50 mT m^{-1} and the frequency encoding gradient to 4 mT m^{-1}. What would be the thickness of the selected slice, and the FOV in the frequency encoding direction?

Answer

This question requires you to distinguish correctly between the two bandwidths associated with slice selection and frequency encoding, and to deal confidently with units. The slice thickness is 1 cm and FOV_{FE} is 11 cm. In the first calculation it is necessary to convert the transmitted bandwidth to units of frequency before using Equation 5.3. In the second, Equation 6.1 is used, and no conversion is necessary, as the equation requires the receiver bandwidth to be expressed in units of frequency.

6.7 Chapter Summary

In this chapter the principles of frequency encoding, and in particular the presence in the signal of several different frequencies, was demonstrated using the Frequency Encoding Demonstrator program. It was shown how each column within a selected slice may be associated with a particular frequency, and that a one-dimensional Fourier transform could be used to extract information about spatial localization from a measured signal. The dependency of the field of view in the frequency encoding direction on gradient strength and receiver bandwidth was described. The addition of frequency encoding means that the signal is now localized to columns within a slice. In the next chapter the last step in the localization procedure will be covered, and the localization of the signal to both row and column in an image will be achieved.

7

Phase Encoding

Phase encoding is the third and final step in the localization of the magnetic resonance (MR) signal. Phase encoding is performed in a direction at right angles to both slice selection and frequency encoding. The *Phase Encoding Demonstrator* program is used early in this chapter to introduce the principles of the technique, and is later used in a more quantitative way.

7.1 Learning Outcomes

When you have worked through this chapter, you should:

- Understand how the MR signal is further localized to a particular row (or column) in the selected slice of the body
- Be able to describe how the values of the parameters that are associated with phase encoding affect the field of view

7.2 Principles of Phase Encoding

In frequency encoding, covered in the previous chapter, the magnetic field gradient was used to ensure that, in the direction of the gradient, each location within the previously selected slice emitted a signal with a different frequency. In phase encoding, which is the topic of this chapter, the magnetic field is used within the slice in a direction at right angles to the frequency encoding gradient. The gradient for phase encoding is applied after the gradient for slice selection and before the gradient for frequency encoding. When the phase encoding gradient is switched on, there is a variation in resonance frequency in the direction of the gradient, just as there is in frequency encoding. The different precessional frequencies mean that phase differences are introduced between the different locations, and when the gradient is switched off, the phase differences remain. These phase differences encode location in the direction of the gradient. Phase encoding differs from frequency encoding in that many different phase encoding gradients are applied in order to encode the whole image.

If the slice selection gradient was applied in the z direction to select a transverse slice, and the frequency encoding gradient in the x direction, then the phase encoding gradient will run in the y direction. We shall use the y direction for the phase encoding gradient throughout this chapter. With this arrangement of gradients, phase encoding will allow determination of the position of the signal in the y direction corresponding with rows in the image.

7.3 Gradient Fields for Phase Encoding

7.3.1 Direction of Gradient for Phase Encoding

The magnetic field gradient for phase encoding is applied in the direction at right angles to both the slice selection and frequency encoding gradients, and at a different time in the pulse sequence. To apply phase encoding from posterior to anterior in a transverse slice across the body, the slice selection gradient is applied running from head to toe, the frequency encoding gradient from side to side, and the phase encoding gradient from back to front. The presence of the gradient means that the precessional frequency is different at each location along the y axis, and so a particular frequency is associated with protons at each y location (Figure 7.1). Several phase encoding gradients are used during an acquisition; the direction of each gradient is the same, but each has a different strength.

7.3.2 Timing of Application of Gradient Field for Phase Encoding

The phase encoding gradient is applied after slice selection and before frequency encoding (Figure 7.2).

7.4 The Phase Encoding Demonstrator

The Phase Encoding Demonstrator program, which is on the CD, can be used to introduce the principles of phase encoding summarized in the first section of this chapter. Start the program as explained in Chapter 1. The interactive demonstrator is shown in Figure 7.3. In the same way as in the Frequency Encoding Demonstrator, a section through the body is

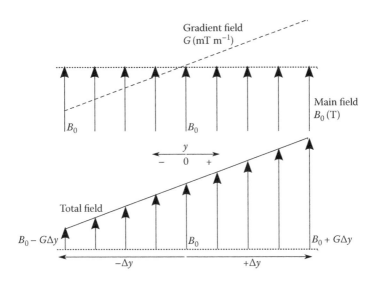

FIGURE 7.1
For phase encoding of a transverse slice the frequency encoding gradient may run in the x direction or the y direction. In Chapter 6 we used an x gradient for frequency encoding, so a y gradient is shown here for phase encoding. The resulting total field varies with location in the y direction, in this case from $B_0 - G\Delta y$ at the left to $B_0 + G\Delta y$ at the right. For a subject lying supine in the scanner, with his body aligned with the z axis, the y direction corresponds with the anterior-posterior direction through his body. The arrows in this figure indicate the size of the main field and not its direction.

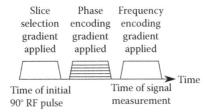

FIGURE 7.2
Each phase encoding gradient field is applied after the radio frequency (RF) pulse and before signal measurement. The repeated application of phase encoding fields, each with a different strength, is indicated by the additional horizontal lines.

shown, with the x axis running from side to side, and the y axis in the anterior-posterior direction. The z axis is at right angles to both x and y, and runs into the screen. This is a conventional arrangement of axes for a horizontal bore magnet, but sometimes the x and y directions might be swapped. Note that the position where $(x, y, z) = 0, 0, 0$ is called the iso-center of the magnet.

Five horizontal lines are shown that represent different values of y (these positions are in different locations in the y direction, in contrast to the frequency encoding demonstrator, where the lines are spaced in the

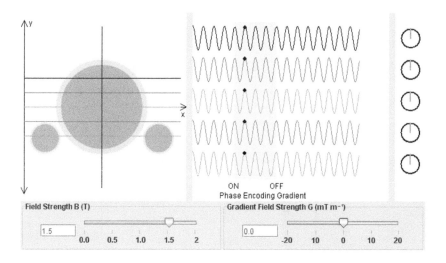

FIGURE 7.3
Screen shot showing the appearance of the Phase Encoding Demonstrator when the program is first run. In the demonstrator the horizontal lines that indicate location are displayed in color (from top to bottom): black, red, orange, green, and blue.

x direction). In exactly the same way as for frequency encoding, each of these lines is associated with a wave representing the frequency of precession at that *y* location. Note, however, that in the phase encoding demonstrator, to reduce the complexity of the diagrams, different amplitudes arising from differing amounts of tissue are not shown. For phase encoding, the period of time for which the gradient is applied is important. So, instead of showing just a few cycles of each signal, a much longer wave train is shown. Time increases from left to right, and in the demonstrator, the time period during which the indicated gradient is applied is illustrated by the yellow band. There are labels indicating when the phase encoding gradient is switched on and off. In the demonstrator the user cannot adjust the timing of the gradient.

When the demonstrator starts up, as shown in Figure 7.3, there is no applied gradient, and the main field strength is 1.5 T. The precessional frequency is the same at all the different *y* locations. For each different location, a black dot is shown on the maximum of the first peak in the yellow region. The dots are all aligned vertically, indicating that with no gradient applied, all the waves are in phase with each other.

7.4.1 Phase Encoding Gradient Fields and Phase Differences

The interactive demonstrator can be used to visualize the effect on the illustrated frequencies of changing the strength of the phase encoding gradient field. Adjust the strength of the gradient by typing 10 into the box (Figure 7.4) and hit the enter key twice. Notice how, within the yellow

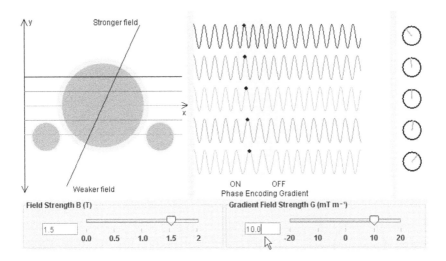

FIGURE 7.4
The Phase Encoding Demonstrator with a gradient of 10 mT m^{-1} selected.

band that indicates when the gradient is switched on, the precessional frequencies at the various locations are now slightly different from each other. The frequency at the center of the body is unchanged, because the gradient field has a value of 0 T at that location. The frequency of precession in the uppermost location is highest because the magnetic field is stronger there than in the other locations. In the demonstrator, the change from weak field to strong field runs in the y direction, and in this case from posterior to anterior through the body.

In phase encoding, the importance of the differences in precessional frequency is that they result in phase differences when the gradient is switched off. In the demonstrator, look at the peaks marked with a black dot. In each case the peak is the first one along after the gradient was applied, but in contrast to the situation when no gradient was applied, they are not in vertical alignment. Use the edge of a piece of paper to join the dots if you are not convinced. When the gradient is switched off, and the precessional frequency returns to f_0 throughout, the phase differences are preserved. So, in the area to the right of the yellow band in the demonstrator, although all the waves again have the same frequency, they remain out of phase.

The far right of the demonstrator includes clock face phase diagrams to illustrate the phase of each signal (Figure 7.5). The phase at the center of the body, where the precessional frequency is unaffected by gradients, is the reference phase, represented by 12 o'clock on the phase diagram. Phase differences introduced by the differing precessional frequencies at other locations are represented by the different locations of the marker in the phase diagrams.

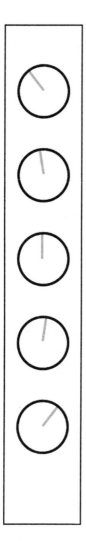

FIGURE 7.5
The clock face diagrams used to indicate phase differences in the Phase Encoding Demonstrator.

Set the gradient to zero again. To demonstrate the effect of the main field on the precessional frequency when no gradient is applied, move the left-hand slider to one side and then the other. This changes the value of the main field strength (B_0). Two extremes are shown in Figure 7.6a and b. Notice how (1) the frequency at all the locations changes as the main field is varied and (2) the phase diagrams all stay the same, meaning that there is no phase shift between the different locations.

Leave the main field fixed at 1.5 T, and use the slider to gradually increase the gradient field strength from 0 mT m^{-1} to 20 mT m^{-1}. Keep your eye on the five colored waves within the yellow band that indicates when the gradient is switched on. Repeat, but this time watch the phase diagrams at the right of the window. Now do the same things again, but using a different value from 1.5 T for the main field. An example using a small main field and strong gradient is Figure 7.6c. You should see that whatever the value chosen for the strength of the main field, the effect of adding a gradient is always the same: the precessional frequency is related to the y location, and there is a resulting y-dependent phase difference. It can be seen in the demonstrator that this phase shift is independent of the main field size, and we shall consider the relationship of the phase shift with other parameters in the next section.

7.5 The Effect of Gradient Strength and Duration on Phase Shift

For a gradient of G_{PE} mT m^{-1} applied in the y direction for a time t_{PE}, it can be shown that the phase difference $\Delta\theta_y$ between the phase at position y and the phase at the center of the field of view (FOV) is given by

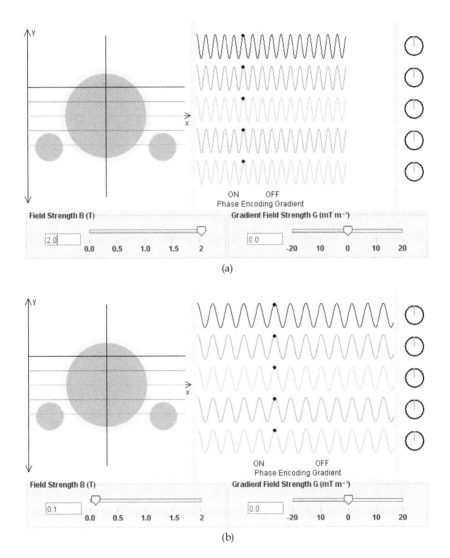

FIGURE 7.6

The effect on precessional frequency of changing the size of the main magnetic field in the Phase Encoding Demonstrator: (a) large main field and no applied gradient, (b) small main field and no applied gradient, (c) small main field and large applied gradient strength.

$$\Delta\theta_y = \gamma y\, G_{PE} t_{PE} \text{ rad} \tag{7.1}$$

or

$$\Delta\theta_y = 360 \cdot \gamma\, y G_{PE} t_{PE} \text{ degrees} \tag{7.2}$$

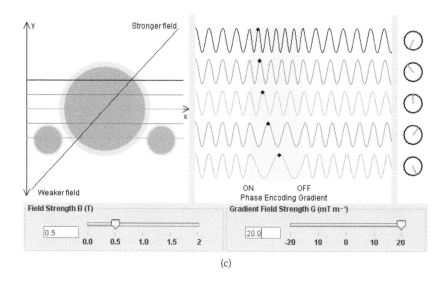

(c)

FIGURE 7.6 (*continued*)

These relationships summarize three properties of the phase difference (i.e., the difference in phase from the phase at the center of the FOV) introduced across the field of view by a gradient:

- The equations do not include B_0, so the phase difference is independent of the size of the main field.
  - In the demonstrator, set the gradient strength to a fixed value. Change the main field strength using the left slider, while observing the clock face indicators of phase difference. The phase difference at each location does not change.
- For a particular gradient applied for a fixed time, the phase difference increases with the displacement y from the center of the field of view.
  - In the demonstrator, set the main field to 1.0 T. Set any non-zero value for the gradient field strength. Note that for all gradient field strengths, the phase difference shown by the clock faces is greatest for the outermost indicators.
- At a particular y location, the phase difference depends on the gradient strength and the time for which the gradient is applied. Larger gradients, or gradients that remain switched on for a long period, lead to a greater phase difference at a particular y location.
  - In the demonstrator, set the main field to 2.0 T and then use the slider to increase the gradient strength slowly. Observe the

uppermost clock face indicator of phase difference. The phase difference increases with increasing gradient strength.

- The time for which the gradient is applied cannot be changed in the demonstrator, but the effect is the same as for changing the strength of the gradient.

TO CLOSE THE PROGRAM

Click on the cross at the top right of the window, and then confirm that you want to end the program.

Worked Example 7.1

A phase encoding gradient is applied in the y direction. The gradient field strength is 0.1 mT m^{-1}, and the gradient is applied for 1.5 ms. Calculate the phase difference, in degrees, for a phase encoding location 30 mm from the isocenter. The gyromagnetic ratio $\gamma = 42.6$ MHz T^{-1}.

The answer is found using Equation 7.2. Substitution of values given in the question leads to the answer: $\Delta\theta_y = 69.0°$.

7.6 Repeated Phase Encoding Steps

Phase encoding is always applied before frequency encoding, and the result of a single measurement made with both phase and frequency encoding applied is that each y position is represented by a unique phase and each x position is represented by a unique frequency. These unique combinations of frequency and phase encode the x and y coordinates of each pixel. Although Fourier analysis can be used to find all the separate frequency components of the measured signal, it is not possible to perform an analysis on a single phase encoding step that will also pull out all the separate phase components. Instead, it is necessary to acquire the signal many times. The same frequency encoding gradient is used each time, but with a different strength of phase encoding gradient. Each separate phase encoding step corresponds with a row of pixels of the image, so the number of phase encoding steps is the same as the number of rows in the image.

7.6.1 Increment in Gradient Field Strength with Phase Encoding Step

The increment in field strength between successive phase encoding gradients is given by

$$\Delta G_{PE} = \frac{\left(G_{PE\,max} - G_{PE\,min}\right)}{N_{steps} - 1} \tag{7.3}$$

where G_{PEmax} and G_{PEmin} are the maximum and minimum available gradients, respectively, and N_{steps} is the required number of phase encoding steps. The value for the gradient strength at each phase encoding step can be found using Equation 7.3 by starting with G_{PEmin} and then repeatedly adding the increment.

Worked Example 7.2

In the demonstrator, the minimum gradient is -20 mT m^{-1} and the maximum gradient is 20 mT m^{-1}. Five phase encoding steps are required; calculate the value of the phase encoding gradient strength at each step.

Substituting for the maximum and minimum gradients and the number of phase encoding steps in Equation 7.3 gives $\Delta G_{PE} = 10$ mT m^{-1}. The first step has a gradient field strength of -20 mT m^{-1}, and the others are found by adding the increment repeatedly: -10 mT m^{-1}, 0 mT m^{-1}, 10 mT m^{-1}, and 20 mT m^{-1}. The image would have five rows. Acquired images are, of course, much larger than in this example, and have perhaps 128 or 256 rows.

Question 7.1

What is the increment in field strength between successive phase encoding gradients for 256 phase encoding steps, where the minimum gradient is -20 mT m^{-1} and the maximum 20 mT m^{-1}?

Answer

From Equation 7.3 the increment in field strength between successive phase encoding gradients is found to be 0.16 mT m^{-1}.

Question 7.2

Prepare a sketch of the left-hand diagram in the demonstrator, showing the body cross section without an applied gradient. Use the demonstrator with $B_0 = 1$ T. Input the first of the five gradient field strengths calculated in Worked Example 7.2. Copy the appearance of the gradient that is shown in the left-hand diagram to your sketch. Repeat for each of the other gradient field strengths, adding the appearance of each to the same sketch.

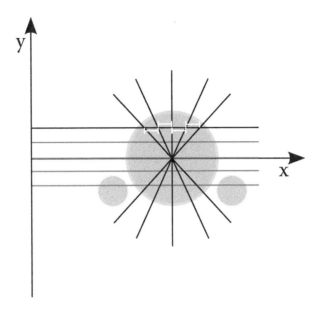

FIGURE 7.7
Feedback for Question 7.2. The white markers indicate how at a particular *y* position, there is a unique increment in the value of the gradient field between successive phase encoding steps.

Answer

Your finished diagram should look like Figure 7.7. Additional white markers are shown in Figure 7.7 to emphasize that at a selected *y* position, there is a unique increment in the value of the gradient field between successive phase encoding steps.

7.6.2 Increment in Phase Shift with Phase Encoding Step

In the previous section we saw that the gradient strength is incremented by the same amount with every step, meaning that the gradient field strength has a linear relationship with phase encoding step. There is also a linear relationship, given in Equation 7.2, between gradient strength and phase shift at a particular *y* location. As a result, the phase shift must also increment by an amount that is the same size on every phase encoding step. This means that at any *y* position, the phase shift introduced by each successive gradient increments by an amount that is unique to that *y* position (Figure 7.8). For example, in one row of the image, the phase shift might change by 30° on every phase encoding step, while in another row farther from the isocenter, the change is always 60°.

The key point here is that it is the *increment in phase shift* between successive phase encoding steps that encodes the *y* location.

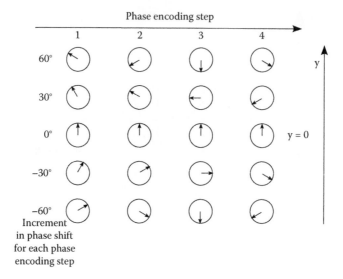

FIGURE 7.8
The increment, between successive phase encoding steps, in the value of the phase shift at any *y* location is unique to that location. This is illustrated here using five rows of the image, and four successive phase encoding steps. For each row (i.e., *y* location) there is a fixed increment in phase shift, so all spins in that row change by that amount on every phase encoding step. The increment is zero for the center row, so there is no change in phase shift for that row.

Question 7.3

This question is very similar to Question 7.2, but instead of considering the gradients at each phase encoding step, we look at the phases instead. Copy the array of clock face diagrams in Figure 7.9. These represent the five phase diagrams in the demonstrator for the five different gradient field strengths calculated in Worked Example 7.2. Use the demonstrator with $B_0 = 1$ T. Input the first of the five gradient field strengths calculated in the worked example. Copy the appearance of the phase diagrams to your copy of the array of clock faces in the correct column for the gradient field strength. Repeat for each of the gradient field strengths.

Answer

As you drew the phase diagrams, you would have noticed that the increment in phase across a row was the same for each gradient. This exercise could be repeated using a different value for the main field and the result would be the same: the increment in phase shift between successive phase encoding steps is constant for any particular *y* location.

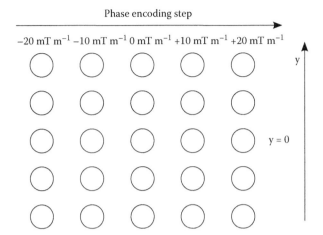

FIGURE 7.9
Diagram to be completed in Question 7.3.

7.6.3 FOV with Multiple Phase Encoding Steps

To ensure that there are no repeated values of phase shift, the phase shift across the whole of the FOV must be no greater than 360° (2π radians). Larger values could not be distinguished from values in the range 0 to 360°, so their presence would lead to image artifacts because there would be ambiguity about the location of the source of the signal. When this condition is applied, it can be shown that the FOV in the phase encoding direction is given by

$$FOV_{PE} = \frac{1}{\gamma \Delta G_{PE} t_{PE}} \tag{7.4}$$

where γ is the gyromagnetic ratio, ΔG_{PE} is the increment in field strength between successive phase encoding gradients, and t_{PE} is the time for which each phase encoding gradient is applied.

It is notable that the FOV in the phase encoding direction does not depend on the absolute value of gradient field strength G_{PE}, but instead is inversely related to ΔG_{PE}, which is the increment in field strength between successive phase encoding gradients. Smaller fields of view are associated with large increments in gradient between phase encoding steps, and with gradients applied for longer times.

Question 7.4

The minimum phase encoding gradient is –20 mT m⁻¹ and the maximum phase encoding gradient is 20 mT m⁻¹. The number of phase encoding steps acquired is 256, and for each step the gradient is applied for 1.0 ms. Calculate the FOV in the phase encoding direction. The gyromagnetic ratio $\gamma = 42.6$ MHz T^{-1}.

Answer

This question involves the combination of two equations. Substitute for ΔG_{PE} from Equation 7.3 in Equation 7.4, then substitute the numerical values given in the question. The answer is that the FOV in the phase encoding direction is 15 cm.

7.6.4 Pixel Size with Multiple Phase Encoding Steps

The number of pixels in the phase encoding direction is the same as the number of phase encoding steps. Thus, the pixel size in the phase encoding direction is given by

$$Pixel\ size_{PE} = \frac{FOV_{PE}}{N_{steps}} \tag{7.5}$$

High spatial resolution corresponds with small pixel size, so for a fixed number of phase encoding steps, higher spatial resolution is expected if the field of view, as given in Equation 7.4, is small. Systems with high maximum gradient strengths are expected to have the highest spatial resolution.

7.7 The Data Matrix

A signal is acquired for each phase encoding step, each time with the same frequency encoding gradient applied. The way in which these signals are saved is illustrated in Figure 7.10. Figure 7.10a shows one signal placed into a matrix whose axes are time and phase encoding step (phase encoding step is also a function of time, because each step is separated by the same amount of time). This particular signal is the one where the phase encoding gradient is zero, so it is centrally placed on the phase encoding step axis. Figure 7.10b indicates how later signals are arranged in the matrix, with one measured signal for each phase encoding step. The figure is highly illustrative, because what is saved in practice is not a series of wiggly lines, but a matrix of digital values generated from the signal data. The value saved at each point in the matrix represents the amplitude and phase of the signal at that location. The matrix may be referred to as the *data matrix, frequency domain representation,* or *k-space*. Whatever name is used, the matrix is a digital version of a set of MR signals, where each signal has been acquired with a different value of the phase encoding gradient.

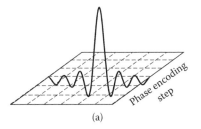 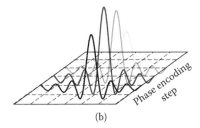

(a)   (b)

FIGURE 7.10

Illustration of the data matrix. (a) The data matrix showing the central positioning of the signal acquired with the phase encoding gradient set to zero. (b) The data matrix with further signals from different phase encoding steps included; each is shown in a different shade of gray. When all the data for an image have been acquired, there will be a signal in the data matrix for every phase encoding step. A matrix of digital values is generated from the signal data that have been arranged in the manner illustrated. The value saved at each point in the data matrix represents the amplitude and phase of the signal at that location.

It is not usual to display the data matrix itself; instead, the matrix is converted, by performing a two-dimensional Fourier transform, into an MR image that maps the locations of signals in the slice. A two-dimensional transform is required because it is performed on a two-dimensional array and not just on a one-dimensional list of numbers. The Fourier transform allows the signals associated with particular frequencies (x locations) and phases (y locations) to be extracted. By performing slice selection, frequency encoding, and phase encoding, followed by Fourier transformation, an MR image has been produced.

DIFFERENT LOCATIONS IN THE DATA MATRIX

The axes of the data matrix are not the same as the x and y axes of the image, but instead represent spatial frequency. Both positive and negative values of spatial frequency are present, with zero placed at the center of the data matrix. The region of low spatial frequency, close to the center of the data matrix, corresponds with features in the image that are represented by low spatial frequency, such as large shapes and slow variations in gray scale. The periphery of the data matrix represents the high spatial frequencies in the image, and is associated with details and edges in the image. Changes to a particular area of the matrix therefore do not affect a particular area in the image; instead, they affect a particular range of spatial frequencies. For example, if an acquisition was performed that omitted filling the periphery of the data matrix, the resulting image would appear blurred compared with one calculated from a completely filled data matrix.

7.8 Additional Self-Assessment Questions

Question 7.5

Which of the following statements are true?

(a) In frequency encoding, location in the frequency encoding direction is encoded by the precessional frequency.

(b) In phase encoding, the phase difference between a pair of locations in the phase encoding direction is always the same.

(c) In phase encoding, at a given location in the phase encoding direction, the phase shift introduced by successive gradients increments by the same amount for each gradient.

(d) In frequency encoding, the frequency encoding gradient is applied at the time of signal measurement.

(e) Phase encoding is performed at the same time as frequency encoding.

Answer

The answers (a), (c), and (d) are correct.

Question 7.6

The minimum phase encoding gradient is -30 mT m^{-1} and the maximum phase encoding gradient is 30 mT m^{-1}. If the increment in field strength between successive phase encoding gradients is 0.3 mT m^{-1}, how many phase encoding steps will there be?

Answer

Use Equation 7.3. The number of phase encoding steps is 201.

Question 7.7

Select the correct statements:

(a) If other variables are fixed, the greater the maximum phase encoding gradient available, the higher the achievable spatial resolution.

(b) Within a single phase encoding step, the phase difference from the phase at the center of the field of view is not affected by the strength of the main field.

(c) The strength of the phase encoding gradient is the same for every phase encoding step.

(d) The strength of the frequency encoding gradient is the same for every phase encoding step.

(e) The strength of the frequency encoding gradient is increased for every phase encoding step.

Answer

The answers (a), (b), and (d) are correct.

Question 7.8

The minimum phase encoding gradient is -10 mT m^{-1} and the maximum phase encoding gradient is 10 mT m^{-1}. If the increment in field strength between successive phase encoding gradients is 0.04 mT m^{-1} and each phase encoding gradient is applied for 3 ms, then what is the expected FOV in the phase encoding direction? The gyromagnetic ratio $\gamma = 42.6$ MHz T^{-1}.

Answer

The FOV is calculated to be 19.6 cm (from Equation 7.4).

Question 7.9

A radiologist desires images where the pixel size is 1 mm or smaller for 128 phase encoding steps. Is this target achievable in the phase encoding direction using a system where the minimum phase encoding gradient is -10 mT m^{-1}, the maximum phase encoding gradient is 10 mT m^{-1}, and each phase encoding gradient is applied for 3 ms? The manufacturer has offered an upgrade package that would increase the maximum gradient to 20 mT m^{-1} and decrease the minimum phase encoding gradient to -20 mT m^{-1}. Would the new gradients improve the spatial resolution? What is the new pixel size in the phase encoding direction for 128 phase encoding steps?

Answer

The pixel size for the original setup can be calculated by first substituting Equations 7.3 and 7.4 into Equation 7.5. The pixel size is calculated to be 0.388 mm, so the original system can achieve the required 1 mm pixel size. Because the difference between the maximum and minimum gradient strengths has doubled, the new gradients would be expected to improve the spatial resolution by reducing the pixel size by a factor of 2.

BACK PROJECTION IN MRI

If you are familiar with the technique of filtered back projection, which is used in x-ray computed tomography (CT) imaging to generate slices from projection data, you may be wondering why it is not used to generate slices in MRI. Early MRI work did in fact use an analogous method. Instead of acquiring projections from different angles through 360°, as is done in CT, sets of data were acquired using gradients at different angles. This technique allowed very small fields of view to be selected, but it was very sensitive to inhomogeneities in the main field and in the gradient, leading to artifacts. In contrast, the Fourier transform approach described in this chapter is relatively insensitive to such inhomogeneities, and has become the standard technique for image generation.

7.9 Chapter Summary

The final step in spatial localization, phase encoding, was described in this chapter. The Phase Encoding Demonstrator was used first to illustrate the general principles of phase encoding, and later to help emphasize how the y location is encoded in the increment in phase shift between successive phase encoding steps. The linear relationship of phase difference from the isocenter with various imaging parameters was used as the basis for a qualitative description of the properties of the phase difference and for numerical work. The data matrix (often called k-space) was introduced. The value saved at each point in the data matrix represents the amplitude and phase of the frequency and phase encoded signal. Information about spatial location is encoded in the amplitude and phase, and this information can be retrieved using Fourier transformation. The preceding three chapters have concerned the methods used to determine the spatial location of an MR signal. In the next chapter these methods will be combined with the use of RF pulses in order to produce an image. The first imaging sequence to be considered is the spin-echo imaging sequence.

8

The Spin-Echo Imaging Sequence

Several topics that have been covered in earlier chapters are now brought together, and we shall see how an image can be generated. Spatial localization, which was described in Chapters 5 to 7, will be combined with operations designed to emphasize differences between the signals emitted by materials with different relaxation times, covered in Chapter 4. When pulse sequences are combined with spatial localization in this way, they are called imaging sequences. The first imaging sequence that we shall consider is the spin-echo sequence. The *Spin-Echo Demonstrator* program is used in this chapter to support the description of the principles of the imaging sequence.

8.1 Learning Outcomes

When you have worked through this chapter, you should:

- Be able to describe the sequence and roles of radio frequency (RF) pulses used in the spin-echo imaging sequence
- Be able to outline the relationships among the echo strength, transverse relaxation time, and time to echo (TE) in the spin-echo imaging sequence

8.2 The Concept of the Spin-Echo Sequence

We saw, when working with the saturation recovery program in Chapter 4, that a large signal would be acquired if it were measured immediately after a 90° RF pulse tips the net magnetization vector into the xy plane. Measurement immediately after the 90° RF pulse corresponds with using a very short TE (time from 90° RF pulse to measurement). In practice, however, imaging sequences based on saturation recovery and a very short TE are not used. The reasons are as follows:

- It is difficult to set up the system electronics to measure a signal immediately after an RF pulse.

- Time is needed to apply the three sets of gradients for spatial localization in the imaging sequence.

- T2* decay can very quickly reduce the size of the signal to be measured.

- The influence of T2 is lost when a very short TE is used, so T2-weighting cannot be achieved with a very short TE.

The spin-echo (SE) sequence is designed to allow the necessary delay in the imaging sequence, while at the same time minimizing the signal-reducing effect of T2* decay. The key point about the magnetic field inhomogeneities that contribute to T2* relaxation is that they are not random. This means that their effect can be cancelled out by introducing an operation to reverse their effect on the spins. In the spin-echo sequence, this is achieved by including a 180° RF pulse in the pulse sequence. In the following section, the way in which this pulse affects the spins is introduced.

8.3 Demonstration of the Principles of the Spin-Echo Sequence

8.3.1 The Spin-Echo Demonstrator

The Spin-Echo Demonstrator program, which is on the CD, is a good way to visualize how the spin-echo sequence works. The demonstrator will show three different vectors, each representing a different spin, in the $x'y'$ plane of the rotating frame of reference. The rotating frame of reference was introduced in Chapter 3, and we saw there that if the axes are rotating at the Larmor frequency in the direction of the curved arrow, any vector precessing at the same frequency (i.e., at the Larmor frequency) will appear to be stationary. Start the spin-echo demonstrator program as explained in Chapter 1, and click on the large *90° Pulse* button at the left of the window so that the diagrams appear. The black vector in the demonstrator always experiences a magnetic field of B_0 and appears stationary in the rotating frame of reference. In the demonstrator the blue vector experiences a field less than B_0, and by moving the blue slider to the left, the difference from B_0 may be increased. Similarly, the orange vector experiences

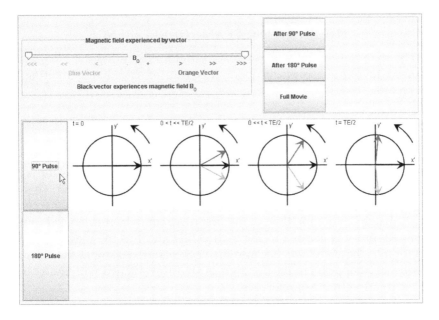

FIGURE 8.1
Screen shot of the Spin-Echo Demonstrator showing the appearance of the display after clicking on the *90° Pulse* button. The parts of the image that appear in orange in the demonstrator are here seen in a dark gray, and blue is represented by a light gray. The orange (dark gray) arrow represents an example of spins precessing at a frequency higher than the Larmor frequency, and the blue (light gray) arrow is an example of spins precessing at a frequency lower than the Larmor frequency.

a field greater than B_0, and by moving the orange slider to the right, the difference from B_0 may be increased.

- The slider on the left is labeled with blue arrows. Move it to the extreme left of its range.
- Move the other slider, which is labeled with orange arrows, to the extreme right of its range.
- Click on the *90° Pulse* button (Figure 8.1).

You should now see a row of four diagrams (Figure 8.1), all of which represent the $x'y'$, or transverse, plane of magnetization in a rotating frame of reference. The row of diagrams illustrates the positions of the three selected spins at increasing times after the 90° RF pulse. Each of the four diagrams represents a different point in time, increasing from $t = 0$ for the leftmost diagram to $t = \mathrm{TE}/2$ for the one at the right.

THE ROTATING FRAME OF REFERENCE

Use the rotating frame of reference program from Chapter 3 if you need to remind yourself about this way of visualizing rotating vectors.

8.3.1.1 $t = 0$

Before the spin-echo sequence starts, the net magnetization vector is in the z direction and there is no magnetization in the $x'y'$ plane. This is the situation before the first of the four diagrams in the demonstrator. Immediately before the time of the first diagram (which is labeled $t = 0$), a 90° RF pulse is applied in the y' direction and tips the magnetization vector into the $x'y'$ plane. Because the vector is precessing at the Larmor frequency, it appears in the diagram as an arrow aligned with the x' axis of the rotating frame of reference.

8.3.1.2 $0 < t \ll TE/2$

Now look at the second diagram along, where a short period of time has elapsed since the 90° RF pulse was applied. In this diagram, the black vector component is accompanied by a blue and an orange arrow. These represent spins that are precessing at frequencies other than the Larmor frequency. The different frequencies arise because, as we know from our understanding of T2 and T2* relaxation, not all the spins experience the same magnetic field strength. Some spins will experience a stronger field and so precess more quickly, and some a weaker field and so precess more slowly. In the demonstrator, spins precessing more quickly than the Larmor frequency are shown in orange, and those that precess more slowly are shown in blue. The direction of precession is anticlockwise (indicated by the curved arrow). This is why, as time goes by, the faster precessing orange arrow is above the x' axis and drawing ahead of the stationary Larmor frequency vector. The blue arrow appears below the x' axis as it falls behind the stationary Larmor frequency vector.

8.3.1.3 $0 \ll t < TE/2$ and $t = TE/2$

As time passes, the orange and blue spins in the diagram spread out away from the black one. This spreading represents the process of dephasing, which arises from the different rates of precession. Overall, the effect of dephasing is to reduce the size of the $x'y'$ component of magnetization, in exactly the same way as we saw when discussing T2 and T2* relaxation

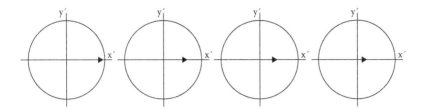

FIGURE 8.2
An illustration of the effect over time of dephasing on the size of the $x'y'$ component of magnetization following a 90° RF pulse.

in Chapter 4. The orange and blue arrows move farther and farther away from the black one as time goes by, illustrating the dephasing that has occurred. As dephasing continues, the size of the $x'y'$ component of the net magnetization vector gets smaller (Figure 8.2).

In Chapter 4 we saw that the two processes that lead to dephasing are:

- Spin-spin interactions, which result in T2 relaxation
- Inhomogeneity of the magnetic field, which leads to the more rapid T2* relaxation

The spin-spin interactions associated with T2 relaxation are usually ignored when explaining the principles of the spin-echo sequence, and we shall ignore the effect in this chapter. Instead, we concentrate purely on the dephasing that arises from inhomogeneities of the magnetic field. These inhomogeneities are in fixed locations, and their sizes do not change with time. In the spin-echo sequence, the initial 90° RF pulse is followed by a 180° rephasing, or refocusing, RF pulse. The 180° RF pulse is applied in the y' direction, a period of time TE/2 after the initial 90° RF pulse. It is this 180° RF pulse that will cause the effects of the inhomogeneities to be cancelled out.

In our demonstrator, the refocusing pulse is applied at the time of the rightmost diagram. The effect of this 180° RF pulse is to flip the arrows through 180°, so that the black arrow that points to the right along the x' axis before the pulse, points to the left after the pulse is applied. It can be a little harder to work out the effect of the pulse on vectors that are not aligned with the x' axis, but one way to visualize the result is to fold or mirror the diagram along the y' axis as indicated in Figure 8.3.

In the demonstrator, the diagram representing a time immediately following the application of a 180° RF pulse will be seen at the left of the second row of images. This row appears when the button *180° Pulse* is clicked (Figure 8.4). Click on the *180° Pulse* button now; you will see that the diagram at the bottom left is the top-right diagram, mirrored about the y' axis.

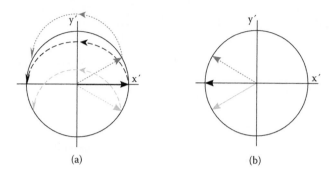

(a)   (b)

FIGURE 8.3

(a) The effect of the 180° RF pulse on out-of-phase spins. (b) The result is to mirror the diagram about the y' axis.

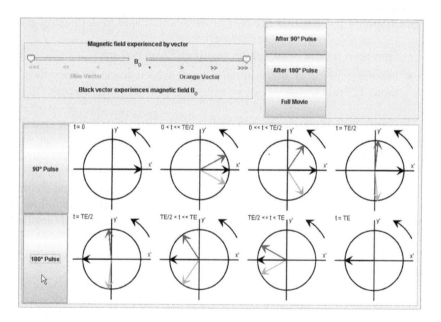

FIGURE 8.4

Screen shot of the Spin-Echo Demonstrator showing the appearance of the display after clicking on the *180° Pulse* button. The parts of the image that appear in orange in the demonstrator are here seen in a dark gray, and blue is represented by a light gray. The orange (dark gray) arrow representing spins precessing at a frequency higher than the Larmor frequency is now behind the Larmor frequency arrow and the blue (light gray) arrow ahead.

8.3.1.4 $t = TE/2$

The effect of the 180° RF pulse on spins rotating at the Larmor frequency is to change the direction of the black vector so that it still appears stationary in the rotating frame of reference, but points to the left in our diagram instead of to the right. The direction of precession is still the same, since the $x'y'$ plane is still perpendicular to the direction of the unchanged main field (the z direction). In the diagram, it now appears that the orange vector, which had drawn ahead of the Larmor frequency vector, is now lagging behind it. The inhomogeneities that caused spins to precess at frequencies higher than the Larmor frequency still do so, but because of the change in direction of the vectors at the 180° RF pulse, the effect of the higher frequency is now to bring the vector back into phase with the reversed Larmor frequency vector. In a similar way, the effect of the 180° RF pulse is to move the blue vector, which was behind the Larmor frequency vector, to be ahead of the reversed Larmor frequency vector. As the spins associated with this vector still precess more slowly, the vectors will gradually come back in phase.

8.3.1.5 TE/2 < t << TE and TE/2 << t < TE

Overall, the effect of the change in direction of the vectors at the 180° RF pulse is to reverse the dephasing. Where previously the orange and blue arrows were drawing away from the black one with time, they now move closer to it.

8.3.1.6 t = TE

At a time TE after the initial 90° RF pulse, all the spins are back in phase. When the spins are in phase the vector sum is large (Figure 8.5) and gives a signal known as the *echo*. In the spin-echo sequence, the signal is measured at a time TE after the 90° RF pulse.

8.3.2 Further Use of the Spin-Echo Demonstrator

8.3.2.1 Adjusting the Size of the Field Inhomogeneities

In the first example, by placing the sliders at the extreme ends of their scales we chose two spins that were experiencing magnetic fields considerably

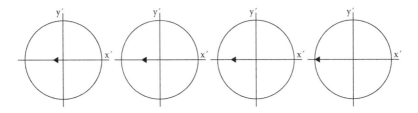

FIGURE 8.5
An illustration of the effect over time of rephasing on the size of the $x'y'$ component of magnetization following a 180° RF pulse.

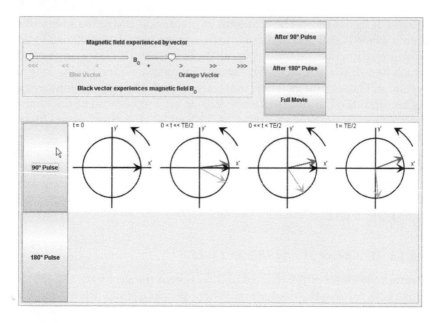

FIGURE 8.6
Screen shot of the Spin-Echo Demonstrator showing the appearance of the display after clicking on the *90° Pulse* button, using different settings for the field inhomogeneities than previously. The parts of the image that appear in orange in the demonstrator are here seen in a dark gray, and blue is represented by a light gray.

different from the main field B_0. By the time $t = $ TE/2 the blue and orange arrows were well separated. The demonstrator will allow you to adjust the size of the magnetic field experienced by the spins associated with the orange and blue vectors. For example:

- Leave the slider for the blue vector at the extreme left of its range.
- Move the orange slider to the left, so that it is about a quarter of the way along the slider.
- Click on the *90° Pulse* button (Figure 8.6). Note that the effects of changes to the sliders are not seen until one of the two buttons is pressed.

In the resulting display we still see the spins spreading out as time goes by, but as the orange spin is experiencing a field only a little greater than B_0, it does not spin much faster than the Larmor frequency, and so does not move far away from the black arrow.

Question 8.1

Select a different pair of values for the blue and orange vectors and generate the top row of diagrams by clicking on the *90° Pulse* button. On a piece of paper, sketch the diagrams you expect to see in the bottom row once the *180° Pulse* button is pressed. Check your predictions using the demonstrator. Repeat with another pair of values.

VARIATIONS IN SPIN-ECHO SEQUENCE
$x'y'$ PLANE DIAGRAMS

You may see variations in the diagrams showing the spin-echo sequence. For example:

- The diagrams may be drawn with clockwise rather than anticlockwise precession. This would mean, for example, that the faster spins would be shown below, instead of above, the x axis.
- The initial 90° RF pulse in our diagram is applied along the y' axis, which aligns the vector along the x' axis. The diagram can also be drawn with the 90° RF pulse applied along the x' axis, which will align the vector along the y' axis.
- Our 180° RF pulse is applied in the y' direction, and this means that the mirroring of the spins takes place about the same y' axis. The 180° RF pulse is occasionally shown applied in the x' direction, in which case the mirroring is done about the x' axis.
- In our diagrams the effect of T2 relaxation is ignored, to emphasize the removal of the T2* effects. Sometimes the diagrams are drawn showing that the resultant vector at the time of measurement is smaller than it was at the beginning of the sequence. The reduction in size is because T2 decay has continued throughout the time taken for the sequence.

8.3.2.2 *Viewing a Movie of Dephasing*

We can use the demonstrator to visualize the dephasing and rephasing in movies. Click on the button at the top right labeled *After 90° Pulse*. This will show you a movie of the period represented by the top row, after the 90° RF pulse but before the rephasing pulse. Six representative vectors are shown dephasing. Figure 8.7 shows the movie part way through. In the movie, spins precessing at the Larmor frequency are shown in black,

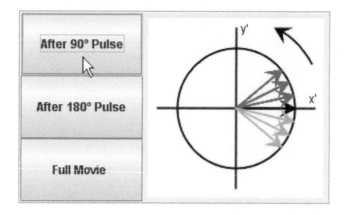

FIGURE 8.7
A still from the movie sequence in the Spin-Echo Demonstrator. The parts of the image that appear in orange in the demonstrator are here seen in a dark gray, and blue is represented by a light gray.

spins precessing at greater than the Larmor frequency are in orange, and spins precessing at less than the Larmor frequency are in blue. The curved arrow shows the direction of precession. This is the case here, but it is best not to make any assumptions about directions and colors when interpreting diagrams like these. Start by looking for an indication of the direction of precession, then decide where a spin precessing at a higher frequency than the Larmor frequency would be seen. For example, if the direction of precession in the movie were clockwise, then the blue arrows below the x' axis would be the fast spins.

Now view the movie for the period after the rephasing pulse using the *After 180° Pulse* button. The button labeled *Full Movie* will show you the dephasing and rephasing parts of the sequence one after the other. Make sure you can identify which of the three orange or blue vectors correspond with the field strengths close to, and very different from, the main field. If you have any doubts, use the sliders in the spin-echo demonstrator to choose the magnetic field, then generate two rows of images to compare with the movies.

8.4 TR and TE

We have been using the abbreviations TR (repetition time) and TE since Chapter 4 when pulse sequences were first introduced. Now, with the spin-echo imaging sequence, the relevance of the abbreviations can be understood.

8.4.1 Repetition Time (TR)

TR represents the repetition time, and it is the time between successive applications of the 90° RF pulse. The pulse needs to be applied repeatedly so that measurements can be made using different values of the phase encoding gradient.

8.4.2 Time to Echo (TE)

TE is the time between the application of the 90° RF pulse and the measurement of the signal. Although the name *TE* is used for the same period in other sequences, it is in the spin-echo sequence that the name *TE* has relevance. TE stands for time to echo or echo time, because the measured signal in the spin-echo sequence is called an echo. Do not forget that the diagrams used to explain the spin-echo sequence deliberately ignore the effect of T2 relaxation. The impression given is that any length of TE could be used, and an equally large echo would be measured for all values of TE. In reality, T2 relaxation continues during the period TE, and the refocused vector will be smaller than the original by an amount predicted by exponential decay with time constant T2. This relationship between TE and the echo strength is shown in Figure 8.8.

Both TR and TE can be chosen by the operator, and we shall see in the next chapter how the choice of values for TR and TE can affect the appearance of the image through T1- and T2-weighting.

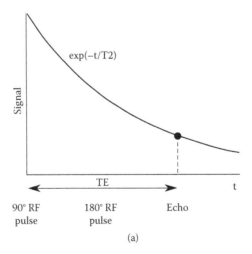

(a)

FIGURE 8.8
Graphs showing how the echo signal size is dependent on TE and T2. (a) The echo is measured TE after the 90° RF pulse, during which time exponential T2 decay occurs (curve). (b) For a tissue with a particular T2, a higher signal is obtained by using a shorter TE. (c) For a fixed TE, a tissue with a longer T2 will give the higher signal.

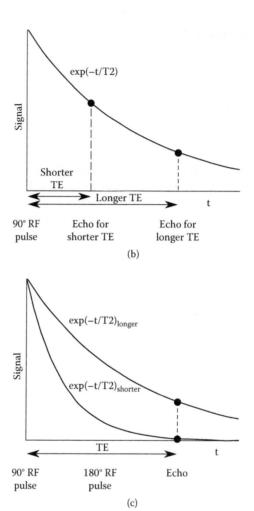

(b)

(c)

FIGURE 8.8 (*continued*)

TO CLOSE THE PROGRAM

Click on the cross at the top right of the window, and then confirm that you want to end the program.

Question 8.2

Figure 8.9a is a diagram showing the state of the spins immediately before the rephasing pulse. Copy the blank diagram in Figure 8.9b and sketch the appearance if an echo time (TE) only half as long were used. Would you expect the signal magnitude of the echo to be greater or less for the shorter TE than the longer TE?

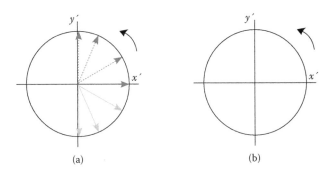

FIGURE 8.9
Diagrams for Question 8.2. The curved arrow shows the direction of precession.

Answer

Your sketch should show the fan of arrows still centered on the x' axis but filling only about 90° instead of 180°. Each of the spins has traveled half as far because only half the time period has gone by. The signal magnitude is expected to be greater for the shorter TE, as there is less time for T2 decay to occur. Remember that the diagrams used to explain the spin-echo sequence are simplified, and T2 relaxation is ignored in the diagrams.

Question 8.3

The spin-echo sequence is used to image a test object made of a material with a very short T2. What difference in echo strength do you expect between acquisitions using a short TE and a long TE?

Answer

If a material of short T2 is imaged with a spin-echo sequence, then a rapid decay of the signal in the $x'y'$ plane is expected. If a long TE is used, it is possible that T2 decay will have destroyed the signal during the time taken to dephase and rephase the spins. A larger signal will result from using a short TE.

Question 8.4

The spin-echo sequence is used to image a test object made of a material with a very long T1. What difference in echo strength do you expect between acquisitions using a short TR and a long TR?

Answer

The material has a long T1. If a short TR is used, then full recovery of the z component of magnetization may not have taken place before the next 90° RF pulse. This means that less magnetization will be tipped into the $x'y'$ plane by succeeding 90° RF pulses, leading to a smaller echo signal. A long TR could allow complete longitudinal recovery to take place at every repetition of the sequence, leading to a higher echo signal than for a short TR.

Questions 8.3 and 8.4 introduce the ideas of *T1-weighting* and *T2-weighting*. TR and TE are chosen to give images where the contrast between materials depends mainly on the T1 values, or the T2 values, of the materials being imaged. We shall be considering T1- and T2-weighting in detail in Chapter 9. There is a third type of weighting called *proton density weighting*. In a proton density-weighted image the contrast between materials depends on the number of protons present, rather than on the T1 or T2 values of the materials.

8.5 Timing Diagram

A full description of the imaging sequence includes information about the gradients used for spatial localization as well as the sequence of RF pulses. The temporal relationship between the RF pulses and gradients is shown in a *timing diagram,* which can alternatively be called a *pulse sequence diagram.* The diagram has five rows: one for the RF pulses, one for each of the slice select, phase encoding, and frequency encoding gradients, and one for the measured signal. A simplified timing diagram for the spin-echo sequence is shown in Figure 8.10.

- In the top row, the 90° RF pulse is shown repeated at an interval of TR. The 180° RF pulse occurs at a time TE/2 after the 90° RF pulse.
- The next row shows the G_{ss}, or slice select, gradient. It is applied at the time of the 90° RF pulse, so that only the spins in that selected slice are saturated and can contribute to the signal.
- The phase encoding gradient is applied once per repetition, with a different strength on each occasion. The diagram indicates the multiple gradients by showing several of the gradients at once, which is why it has a striped appearance on the timing diagram.
- The frequency encoding, or readout, gradient is timed to coincide with the time of the echo.
- The last row shows when the signal is measured, at a time TE after the 90° RF pulse.

The diagram of Figure 8.10 is simplified, because it shows the gradients being switched on and off in a way that will permit spatial localization, but without concern regarding the gradient direction or its other effects. In practice, applying a gradient will always cause spins to dephase, because

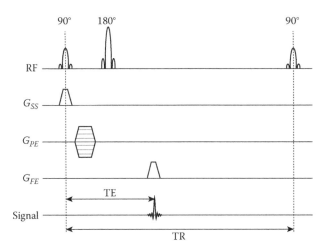

FIGURE 8.10
Simplified timing diagram for the spin-echo imaging pulse sequence.

of the field variation introduced by the gradient. So although the gradient permits signal localization to be performed, the dephasing introduced by the gradient could destroy a signal. So, in real timing diagrams it is usual to see the gradient being applied for a period in the opposite direction from the way in which it is applied for localization. This corrects for the gradient-induced dephasing, but does not change the effect of the gradient for localization purposes.

8.6 Additional Self-Assessment Questions

Question 8.5

The concept of dephasing is very important for understanding the spin-echo sequence. Which of the following statements are true?

(a) Dephasing arises from spin-spin interactions.
(b) Dephasing arises from inhomogeneity of the main magnetic field.
(c) Dephasing takes place in the $x'z'$ plane and accounts for T1 relaxation.
(d) Dephasing takes place in the $x'y'$ plane and accounts for T2 relaxation.

Answer

The true statements are (a), (b), and (d). Remember that both (a) and (b) are true, and that they are the two principal reasons for dephasing.

Question 8.6

Select the correct statements:

(a) T2 decay depends on the homogeneity of the main magnetic field.
(b) T2 decay depends on spin-spin interactions.
(c) Both (a) and (b) are true.
(d) T2* decay depends on the homogeneity of the main magnetic field.
(e) T2* decay depends on spin-spin interactions.
(f) Both (d) and (e) are true.

Answer

T2 decay depends primarily on spin-spin interactions, while T2* decay, which is always more rapid, depends on both the homogeneity of the main magnetic field and spin-spin interactions. So (b) and (f) are the correct statements.

Question 8.7

Which of the following statements are true?

(a) The spin-echo pulse sequence eliminates dephasing due to magnetic field inhomogeneities.
(b) The spin-echo pulse sequence eliminates dephasing due to spin-spin interactions.
(c) The spin-echo pulse sequence eliminates dephasing using a refocusing 180° RF pulse.
(d) The spin-echo pulse sequence eliminates dephasing using a refocusing 90° RF pulse.
(e) The spin-echo pulse sequence eliminates dephasing using a refocusing 180° gradient pulse.

Answer

The true statements are (a) and (c).

Question 8.8

Why does the 180° RF pulse not eliminate the dephasing caused by spin-spin interactions?

Answer

Spin-spin interactions fluctuate randomly, which means that the fluctuations from spin-spin interactions are not identical before and after the 180° refocusing RF pulse. The pulse therefore has no effect on this kind of dephasing. In contrast, the fixed nature of the magnetic field inhomogeneities means that their effects can be cancelled out with the refocusing pulse.

Question 8.9

Select the true statement from the following about the spin-echo sequence:

(a) TE is the time period between successive 90° RF pulses.
(b) TE is the time between the 90° RF pulse and the 180° RF pulse.
(c) TE is a property of tissue that determines how quickly saturation recovery takes place.
(d) TE is the time between the 90° RF pulse and the time at which the echo occurs.
(e) TE is a property of tissue that has an effect on the size of the measured echo.

Answer

The true statement is (d). Now try the next question, which will help to explain why the other choices are incorrect.

Question 8.10

Replace TE in the incorrect statements of Question 8.9 with the parameter that will make the statements true.

Answer

The corrected statements about the spin-echo sequence, which are now all true, are as follows:

(a) TR is the time period between successive 90° RF pulses.
(b) TE/2 is the time between the 90° RF pulse and the 180° RF pulse.
(c) T1 is a property of tissue that determines how quickly saturation recovery takes place.
(d) TE is the time between the 90° RF pulse and the time at which the echo occurs.
(e) T2 is a property of tissue that has an effect on the size of the measured echo.

Question 8.11

The diagram in Figure 8.11 is typical of those that are used to illustrate the spin-echo sequence. It uses a rotating frame of reference and indicates the direction of precession. Three magnetization vectors are shown in the $x'y'$ plane some time after the application of a 90° RF pulse. Note carefully the arrow showing the direction of precession, then label the three magnetization vectors diagram with the three correct labels, chosen from this list:

(a) Experiencing magnetic field $= B_0$
(b) Experiencing magnetic field $> B_0$
(c) Experiencing magnetic field $< B_0$

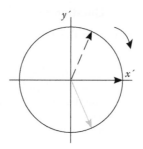

FIGURE 8.11
Figure for Question 8.11. The curved arrow shows the direction of precession.

(d) Experiencing magnetic field $= 0$
(e) Experiencing magnetic field > 0
(f) Experiencing magnetic field < 0

Answer

The rotating frame of reference rotates at the Larmor frequency associated with the main field B_0, so any vector precessing at that frequency appears to be stationary and stays in the same place as time goes by. The middle vector is aligned with the x' axis and should be labeled with statement (a). Any vector experiencing a higher field will precess at a higher frequency, and so appears to draw ahead (in the direction of precession) of vectors precessing at the Larmor frequency. Similarly, those with a slightly lower precessional frequency from a slightly smaller magnetic field get left behind. So the gray vector should be labeled (b) and the dashed vector (c). As always, remember diagrams of this kind consider only the main field inhomogeneities that can be compensated for by using a rephasing pulse. There is also dephasing going on at the same time from spin-spin interactions, which is ignored in this representation.

Question 8.12

Which of the diagrams in Figure 8.12b to d is the correct one to illustrate the effect on diagram Figure 8.12a of a 180° rephasing RF pulse applied in the y' direction?

Answer

The correct diagram is in Figure 8.12c. A common mistake is to choose the diagram in Figure 8.12b, in which each individual arrow has been flipped by 180° about an axis perpendicular to itself, rather than about the y' axis. This topic was covered in Section 8.3.1.

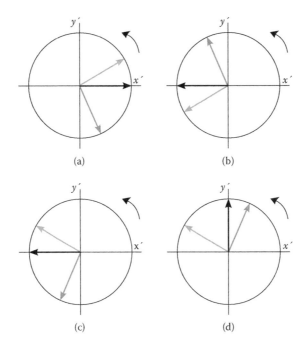

FIGURE 8.12
Figure for Question 8.12. The curved arrow shows the direction of precession.

8.7 Chapter Summary

In this chapter the principles of the spin-echo sequence were described. In particular, the concept of using a 180° rephasing RF pulse was explained, and familiarity was gained with the diagrams commonly used to illustrate the dephasing and rephasing of spins affected by inhomogeneity. The timing diagram, which shows the temporal sequence of the components of a magnetic resonance imaging sequence, was introduced using a simplified version for the spin-echo sequence.

It was noted that the choice of TR and TE for the spin-echo sequence can lead to images that depend strongly on either T1 or T2 values. Such images are described as being either T1-weighted or T2-weighted, and the topic of weighting will be covered in the next chapter. In the next chapter we shall also consider the effect on spatial resolution and image quality of the values of some of the other parameters associated with the spin-echo sequence.

FIGURE 8.28

9

Scan Parameters for the Spin-Echo Imaging Sequence

The choice of acquisition parameters made when using a spin-echo imaging sequence affects the gray-scale characteristics of the image (such as the contrast between different tissues), the spatial or geometrical characteristics of the image (such as pixel size), the noise properties of the image (how grainy it appears), and the time taken to acquire the data. We have already seen that the choice of time to echo (TE) and repetition time (TR) has a major effect on the image contrast, and leads to the possibility of acquiring T1-weighted and T2-weighted images. Image weighting is addressed in more depth in this chapter, and then the influence on the image of other parameters is introduced. The Saturation Recovery program is again used in this chapter, together with the *Image Simulator–Spin-Echo* program.

9.1 Learning Outcomes

When you have worked through this chapter, you should:

- Be confident in predicting the gray-scale appearance of a spin-echo image from knowledge of the values chosen for TR and TE
- Understand the relationship between the spatial characteristics of a spin-echo image and the other acquisition parameters

9.2 Image Gray-Scale Characteristics

We saw in Chapter 8 how the values of TE and TR can influence the relative brightness of different materials in the image. Two approaches to recalling typical appearances of T1- and T2-weighted images for different combinations of TE and TR are included in this section. We look first at rules that may be learned, and then at the use of sketch graphs. Sketch graphs may be used as a memory aid, and provide more flexibility for thinking about different situations.

9.2.1 T1-Weighting, T2-Weighting, and Proton Density Weighting Summarized

9.2.1.1 *T1-Weighting*

In a T1-weighted image (Figure 9.1a), tissues with short T1 appear bright and those with long T1 appear dark. The influence of the T2 of the tissue is minimized.

For T1-weighting:

- TR short (e.g., 250 to 700 ms [3])
- TE short (e.g., 10 to 25 ms [3])

T1-weighted images demonstrate anatomy well and are used when a clear image of structure is required.

9.2.1.2 *T2-Weighting*

In a T2-weighted image (Figure 9.1b), tissues with long T2 appear bright and those with short T2 appear dark. The influence of the T1 of the tissue is minimized.

For T2-weighting:

- TR long (e.g., >2,000 ms [3])
- TE long (e.g., >60 ms [3])

Long T2 values, which appear bright in a T2-weighted image, are associated with increased water content and pathology. T2-weighted images are used to demonstrate pathology.

9.2.1.3 *Proton Density Weighting*

In a proton density-weighted image (Figure 9.1c), tissues with high proton density appear bright and those with low proton density appear dark. The influence of both the T1 and T2 of the tissue is minimized.

For proton density weighting:

- TR long (e.g., >2,000 ms [3])
- TE short (e.g., 10 to 25 ms [3])

In proton density-weighted images, the brightness is related primarily to the number of protons per unit volume, rather than differences in T1 or T2.

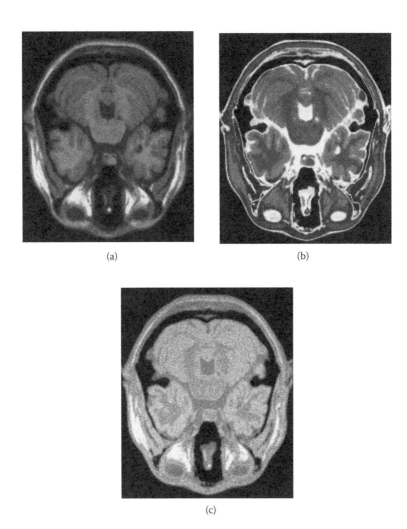

(a)   (b)

(c)

FIGURE 9.1
Examples of images generated using a spin-echo imaging pulse sequence. (a) T1-weighted image with TR = 500 ms and TE = 15 ms, (b) T2-weighted image with TR = 2,500 ms and TE = 90 ms, (c) proton density-weighted image with TR = 2,500 ms and TE = 15 ms. (Image data from the BrainWeb Simulated Brain Database [1,2]. With permission.)

High-water-content tissue, such as brain tissue, has higher proton density than, for example, cortical bone, which contains less water and consequently fewer hydrogen nuclei (protons). Soft tissue contrast is generally less good than T1-weighted and T2-weighted images, so proton density weighting is used less often. Note that although it is possible to reduce the effect of T1 or T2 differences from the choice of TR and TE, proton density weighting is present to some extent in images using all three types of weighting.

9.2.2 Drawing Sketch Graphs

If you learn how to draw just three sketch graphs, then you can use these in a whole range of situations to help you decide which material will appear bright or dark in an image.

The graphs are essentially the same as those you used in the saturation recovery exercises. They represent recovery in the z direction after a 90° radio frequency (RF) pulse (we will call this *sketch 1*) and the signal in the $x'y'$ plane after a further 90° RF pulse (*sketch 2* and *sketch 3*). It is valid to use the saturation recovery plot for the spin-echo sequence even though the plot does not include the dephasing time, 180° flip, and rephasing time of the spin-echo sequence. The saturation recovery plot considers only T2 decay, without the additional effects that lead to the more rapid T2* decay. As the effect of the 180° rephasing pulse of the spin-echo sequence is to remove the additional signal decay and leave just the T2 decay, we can say that the $x'y'$ signal in the saturation recovery plot also represents the signal available from the spin-echo sequence.

9.2.2.1 Sketch 1

Sketch 1 (Figure 9.2) is a plot representing longitudinal relaxation. In this example two materials are indicated, one with a short T1 and one with a long T1. Depending on the question you are addressing, you may need to add other curves, to represent materials with different combinations of T1 and T2. The sketch should also include typical time points to represent short and long TRs; these are indicated by vertical dashed lines in Figure 9.2. The two TR values are relevant for the next two sketches.

INCLUDING PROTON DENSITY IN THE SKETCHES

Note that here we assume that the proton density is the same for all materials so that they all recover to the same plateau level at large values of time. If you want to avoid this assumption, you can include proton density by assigning materials different plateau levels in sketch 1.

9.2.2.2 Sketch 2

The second and third sketch graphs both show transverse relaxation for materials with differing values of T2, but the two graphs are for different values of TR. Sketch 2 (Figure 9.3) is for the long TR case, and sketch 3 for the short TR case. Refer back to sketch 1 (Figure 9.2) and note that the reason both curves in sketch 2 start at the same high value of M_{xy} is because for a long TR, all materials will have completed longitudinal relaxation.

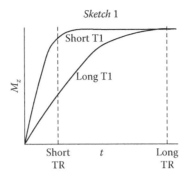

FIGURE 9.2
Sketch 1, which shows longitudinal relaxation for materials with long and short T1 values. The positions for long and short TRs are also shown.

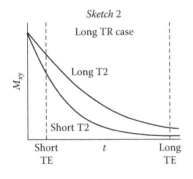

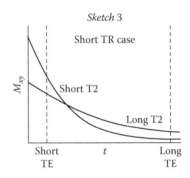

FIGURE 9.3
Sketch 2, which shows transverse relaxation for materials with long and short T2 values after a long TR period.

FIGURE 9.4
Sketch 3, which shows transverse relaxation for materials with long and short T2 values after a short TR period.

Just as we marked long and short TRs on sketch 1, sketch 2 has long and short TE values indicated.

9.2.2.3 Sketch 3

Sketch 3 (Figure 9.4) shows transverse relaxation when TR (in sketch 1) was short. If you refer back to sketch 1 (Figure 9.2), note how for a short TR, the material with a short T1 has almost completed relaxation, but the material with the longer T1 has not. This means that in the second case there is less magnetization available to be tipped into the $x'y'$ plane, and hence the two materials have different starting values of M_{xy}. If you are unsure if a value of a time constant should be considered short or long, remember that if a material has a time constant τ ms, it takes τ ms for the signal to recover to 0.63 of its initial value in the z direction, or in the $x'y'$

TABLE 9.1

T1 and T2 Values for
Worked Example 9.1

Material	T1/ms	T2/ms
A	300	80
B	1,500	200

TABLE 9.2

Combinations of TR and TE
for Worked Example 9.1

TR	TE
Short	Short
Short	Long
Long	Short
Long	Long

plane, to drop to 0.37 of its starting value. Full recovery can be considered to have taken place after a period of about five time constants.

Worked Example 9.1

A specially made test object is used. The materials have T1 and T2 values as shown in Table 9.1. Use sketch graphs to answer the following questions:

(a) Will an image that emphasizes differences in the T1 values of the materials be acquired with a long or short TR?

(b) If a long TR sequence is used, which material do you expect to give the higher signal at short TE?

(c) If a long TR sequence is used, which material do you expect to give the higher signal at long TE?

(d) Which of the combinations shown in Table 9.2 would lead to material A giving a higher signal than material B?

Start by drawing sketch 1 (Figure 9.5a). Label the two curves with A and B as appropriate. Part (a) can be answered from this sketch alone, in which you can see that differences in T1 will only be apparent if a short TR is used. To answer part (b), draw sketch 2 for the long TR case (Figure 9.5b). Label the two curves with A and B as appropriate. From this sketch it can be seen that the long T2 material gives the higher signal at short TE. The same sketch can be used to answer part (c); the signal is higher for the long T2 material at long TE too. Part (d) asks us to find the combination of parameters that leads to material A giving a higher signal than material B. Material A does not have a higher signal in sketch 2, so it is time to draw sketch 3 (Figure 9.5c). From sketch 3 it is clear that the required combination is short TR and short TE.

Worked Example 9.2

Table 9.3 shows T1 and T2 values for three different tissues. Assume that the proton density is the same for each tissue.

(a) We wish to image in such a way that the signal from edema is brighter than, and can be distinguished from, those of white matter and cerebrospinal fluid (CSF). Use sketch graphs to describe at what time point the signal should be acquired if a short TR is used.

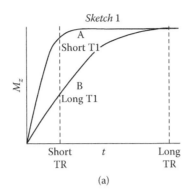

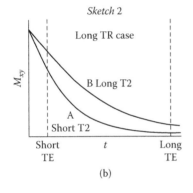

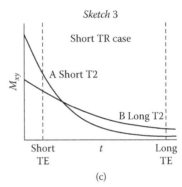

FIGURE 9.5
Sketches for Worked Example 9.1. The T1 and T2 values for materials A and B are shown in Table 9.1. (a) Sketch 1, (b) sketch 2, (c) sketch 3.

TABLE 9.3

T1 and T2 Values for Worked Example 9.2

	T1/ms	T2/ms
White matter	510	67
Edema	900	126
CSF	2,650	180

Source: From Stark, D. D. and Bradley, W. G., Eds., *Magnetic Resonance Imaging*, 3rd ed., Mosby, St. Louis, 1998, 44. (Copyright Elsevier. With permission.)

(b) Use the Saturation Recovery program (previously used in Chapter 4). Select the white matter, CSF, and tissue A tick boxes. For edema, set the *T1 and T2 of your choice* for tissue A to the values shown for edema in Table 9.3. Set *Time to 90 degree flip* (TR) to 400 ms. Increase the *Time from flip to measurement* (TE) until you can see the part of the curve that corresponds with the TE you suggested in (a). Read a value from the plot for a suitable TE.

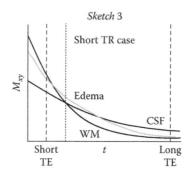

FIGURE 9.6
Sketch 3 for Worked Example 9.2. The T1 and T2 values for the three tissues are shown in
Table 9.3. Edema is shown in gray, and the dotted line shows the time at which the white
matter (WM) and CSF curves (black) have the same value.

Part (a) requires sketch 3, which shows M_{xy} for short TR. From Table 9.3 you
can observe that T2 is short for white matter, long for CSF, and somewhere in
between for edema. Draw sketch 3, adding labels to indicate that the short T2
curve is for white matter and the long T2 curve for CSF. Then add a curve for
the edema signal, which has a decay time between the other two (Figure 9.6).
There is a range of times at which the edema curve represents a higher signal
than those from the other tissues, but the easiest to describe is the time at
which the curves for white matter and CSF cross (that is, have the same value),
and the edema curve lies above them. This time is indicated with a vertical
dotted line in Figure 9.6.

For part (b), a value of TE = 200 ms is suitable to ensure that the crossing
point of the white matter and CSF curves can be seen (Figure 9.7). From this
plot we see that the crossover point between white matter and CSF (orange
and red) is approximately $t = 550$ ms. This corresponds with a TE of 150 ms.
At that point the edema curve lies above both white matter and CSF curves,
as predicted.

Question 9.1

The image shown in Figure 9.8 represents a slice through a test object,
which consists of a cylindrical object set within a block of another
material. The material of the cylinder has both a shorter T1 and a
shorter T2 than the values for the surrounding material. Decide, with
the help of sketch graphs, if the cylinder or the surroundings would
have the higher signal intensity:

(a) In a T1-weighted spin-echo image
(b) In a T2-weighted spin-echo image

Answer

The answers are (a) the cylinder and (b) the surroundings. You should
have started by drawing sketch 1 and labeling the two curves. Sketch 1
should have reminded you that a T1-weighted image has a short TR

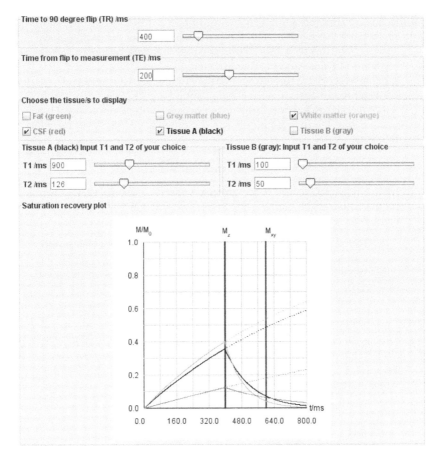

FIGURE 9.7
A screen shot showing the Saturation Recovery program being used to answer Worked Example 9.2.

FIGURE 9.8
Schematic drawing of a slice through a test object, for Question 9.1. The material of the cylinder, shown here in gray, has both a shorter T1 and a shorter T2 than the values for the surrounding material, shown in white.

to ensure differences in T1 have an impact on the signal intensity, so sketch 3 for the short TR case is the relevant plot for part (a). In a T1-weighted image, a short TE is used so that differences in signal are from T1 values and not due to differences in T2. From sketch 3 we see that at short TE, the cylinder has the higher signal intensity. For a T2-weighted image, we need sketch 2 for a long TR. From this graph we expect the difference due to differences in T2 to be emphasized by using a long TE. In this case the surroundings have the higher signal intensity.

Question 9.2

In the image in Figure 9.9, CSF, which has long T1 and long T2, appears brighter than white matter, which has shorter values for both parameters. A long TE was used.

(a) Use sketch graphs to decide if a short TR or a long TR would lead to this result.
(b) What happens to the contrast between the materials if the alternative TR is used?

Answer

Once again, this is tackled by drawing a set of sketches. It can be concluded that for CSF to appear brighter than white matter when a long TE is used, a long TR is required. If a short TR were used (sketch 3), white matter would be brighter than CSF.

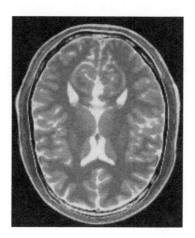

FIGURE 9.9
Image for Question 9.2. (Image data from the BrainWeb Simulated Brain Database [1,2]. With permission.)

Question 9.3

Imagine that you have forgotten what combinations of short and long TR and TE lead to T1-weighted, T2-weighted, or proton density-weighted images. Prepare a table with T1-weighting, T2-weighting, and proton density weighting as the row headings, and with the column headings TR and TE. Complete the table, using sketch graphs to work out if each cell entry should read long or short.

Answer

Your table should appear as Table 9.4. To reach your conclusions, start by drawing all three sketches. You determine that T1-weighting can only be achieved for short TR, because at longer TR the signals from materials with different T1 values are the same. This also means that a long TR will give T2-weighting or proton

TABLE 9.4

Feedback on Question 9.3

	TR	TE
T1-weighted image	Short	Short
T2-weighted image	Long	Long
Proton density-weighted image	Long	Short

density weighting, as differences in T1 will have little effect in this case. Now consider T1-weighting further, using sketch 3, which is the short TR case necessary for T1-weighting. To maintain T1-weighting, a short TE is necessary; otherwise, the effect of T2 will reduce the influence of the different starting values caused by differences in T1. Now use sketch 2, which is the long TR case for T2-weighting or proton density weighting. Here a long TE would better emphasize differences in T2, and so is associated with T2-weighting, while a short TE results in a proton density-weighted image.

9.2.3 Contrast Agents and T1- and T2-Weighting

Contrast agents were introduced briefly in Chapter 4. They can be used to help differentiate different regions of tissue by making the signal higher or lower than it would otherwise have been. Some agents are delivered orally, others intravenously.

9.2.3.1 T1-Shortening Agents

Contrast agents based on gadolinium are the most commonly used. These agents shorten the T1 of tissues where it is present, and so in a T1-weighted image these tissues appear brighter than they would have otherwise. Tumors, for example, may be highly vascularized, and so take up more of the agent than the surrounding tissue. Another property of gadolinium agents is that they can cross the blood–brain barrier if it is compromised, for example, in the presence of intracranial lesions. Gadolinium is non-specific, and is distributed throughout the extracellular space in the body. Other T1-shortening agents, such as some designed for use in the liver, are taken up by healthy tissue. This means that in a T1-weighted image for this type of agent, lesions appear dark against the enhanced background of healthy tissue.

9.2.3.2 T2-Shortening Agents

Gadolinium agents shorten T2 in tissue as well as T1, but the T1 shortening dominates. For other contrast agents, T2 shortening is dominant, which means that the signal on a T2-weighted image is reduced. Iron-based agents (SPIO—superparamagnetic iron oxide) are examples of T2-shortening agents. In the liver, for example, the contrast agent accumulates in normal tissue, so in a T2-weighted image the signal from healthy tissue is suppressed and lesions appear brighter as they have not taken up the contrast agent.

9.2.4 Additional Self-Assessment Questions

Question 9.4

Assign the correct description from this list to each of the three diagrams in Figure 9.10, which illustrate saturation recovery:

- T1-weighting
- T2-weighting
- Proton density weighting

Answer

The order of the diagrams is proton density weighting, T1-weighting, and T2-weighting. The feedback from the next question is also relevant here.

Question 9.5

Assign the correct description from this list to each of the three images in Figure 9.11, which are images of the brain acquired using the spin-echo sequence:

- T1-weighting
- T2-weighting
- Proton density weighting

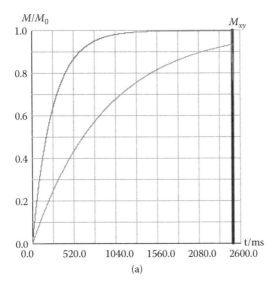

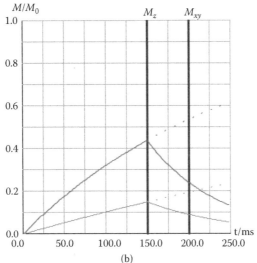

FIGURE 9.10
Diagrams for Question 9.4, generated using the Saturation Recovery program.

Answer

The order of the images is the same as in the preceding question: proton density weighting, T1-weighting, and T2-weighting. You can deduce this because in the proton density-weighted image it can be seen that T1 and T2 differences are minimized. In the T1-weighted

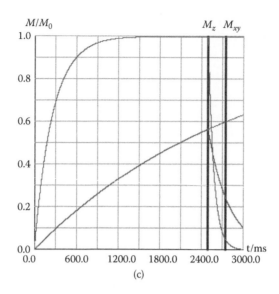

(c)

FIGURE 9.10 *(continued)*

image, fat, which has a short T1, appears bright. In the T2-weighted image, CSF, which has a long T2, appears bright.

Question 9.6

Figure 9.12a shows an image, and Figure 9.12b,c,d has plots of saturation recovery. Which of the saturation recovery plots is consistent with the image?

Answer

Figure 9.12a is a T1-weighted image, in which materials with short T1 appear bright, such as the fat behind the eyes. T1-weighting corresponds with an acquisition with a short TR and a short TE, so the correct answer is (c).

Question 9.7

A saturation recovery diagram is shown in Figure 9.13a; by deducing if the TR and TE shown are long or short, select the matching image from Figure 9.13b,c,d.

Answer

The saturation recovery plot in Figure 9.13a indicates a long TR, long TE sequence. This can be inferred because (1) almost full T1 relaxation has taken place for several of the curves, and (2) there has been time for T2 differences to manifest themselves after the 90° RF pulse. The

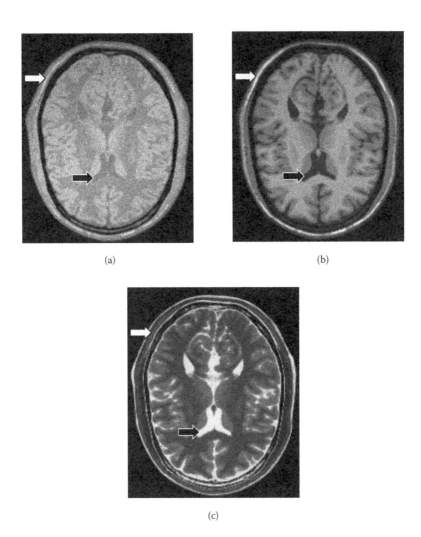

(a) (b)

(c)

FIGURE 9.11
Images for Question 9.5. The white arrows indicate fat; the black arrows indicate CSF. (Image data from the BrainWeb Simulated Brain Database [1,2]. With permission.)

image acquired using such values will be T2-weighted, and the corresponding image is Figure 9.13d, where CSF (with long T2) is bright.

Question 9.8

Assign the correct description from this list to each of the three images in Figure 9.14, which were acquired using the spin-echo sequence:

- T1-weighting
- T2-weighting
- Proton density weighting

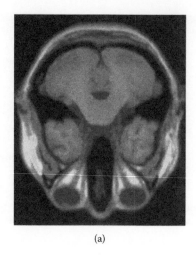

(a)

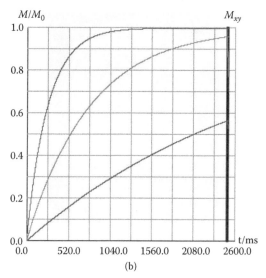

(b)

FIGURE 9.12

(a) Image and (b to d) saturation recovery plots for Question 9.6, generated using the Saturation Recovery program. (Image data from the BrainWeb Simulated Brain Database [1,2]. With permission.)

Answer

Figure 9.14a is a T2-weighted image; for example, look at the vitreous fluid in the eye, which has a long T2 and appears bright. Figure 9.14b is a proton density-weighted image in which T1 and T2 differences

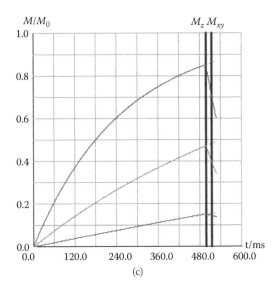

(c)

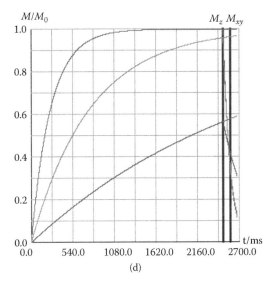

(d)

FIGURE 9.12 *(continued)*

are minimized, and some regions that are well differentiated in the T2-weighted image (Figure 9.14a) are not well differentiated in Figure 9.14b. Figure 9.14c has T1-weighting. Fat, which has a short T1, appears bright, while fluid appears dark.

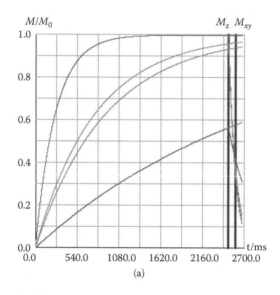

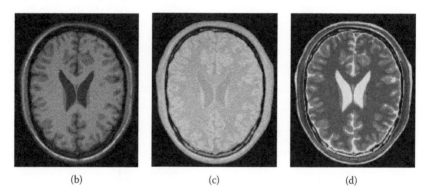

FIGURE 9.13

(a) Saturation recovery diagram and (b to d) images for Question 9.7. (Image data from the BrainWeb Simulated Brain Database [1,2]. With permission.)

Question 9.9

Assign the correct description from this list to each of the three images in Figure 9.15, which illustrate images acquired using the spin-echo sequence:

- T1-weighting
- T2-weighting
- Proton density weighting

Answer

Figure 9.15a,b show T1-weighted images, and Figure 9.15c is T2-weighted.

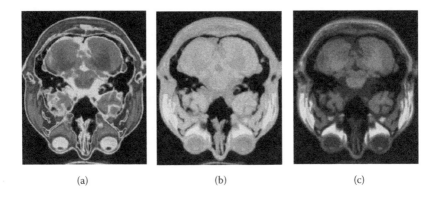

(a)   (b)   (c)

FIGURE 9.14
Images for Question 9.8. (Image data from the BrainWeb Simulated Brain Database [1,2]. With permission.)

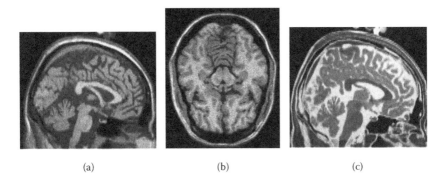

(a)   (b)   (c)

FIGURE 9.15
Images for Question 9.9. (Image data from the BrainWeb Simulated Brain Database [1,2]. With permission.)

9.3 Image Spatial Characteristics

The scan parameters that are responsible for spatial image characteristics, such as the pixel size and field of view (FOV), are related to the RF pulse and the gradients; they are not associated with the timing parameters of the spin-echo sequence itself. The relevant equations associated with spatial image characteristics from Chapters 5 to 7 are summarized in Table 9.5.

The links among the field of view, image matrix size, and spatial resolution were described in Section 2.4.2. Note, however, that using small pixels or thin slices does not guarantee that close objects will be clearly

TABLE 9.5

Summary of Equations Associated with Spatial Image Characteristics

Spatial Image Characteristic	Equation	Equation Number	
Pixel size in slice select direction (i.e., slice thickness)	$Pixel\ size_{SS} = \dfrac{TBW}{G_{SS}\gamma}$	5.3	TBW is the transmitted band-width in units of frequency, G_{SS} is the gradient strength in T m^{-1}, and γ is the gyromagnetic ratio.
Field of view in frequency encoding direction	$FOV_{FE} = \dfrac{RBW}{G_{FE}\gamma}$	6.1	RBW is the receiver bandwidth in units of frequency, G_{FE} is the frequency encoding gradient strength in T m^{-1}, and γ is the gyromagnetic ratio.
Pixel size in frequency encoding direction	$Pixel\ size_{FE} = \dfrac{FOV_{FE}}{N_{FE}}$	6.2	N_{FE} is the number of pixels (matrix size) in the frequency encoding direction.
Field of view in phase encoding direction	$FOV_{PE} = \dfrac{1}{\gamma \Delta G_{PE}\, t_{PE}}$	7.4	The phase encoding gradient is applied for a time t_{PE}, ΔG_{PE} is the increment in field strength between successive phase encoding gradients in T m^{-1}, and γ is the gyromagnetic ratio.
Pixel size in phase encoding direction	$Pixel\ size_{PE} = \dfrac{FOV_{PE}}{N_{PE}}$	7.5	N_{PE} is the number of pixels (matrix size) in the phase encoding direction, and is equal to the number of phase encoding steps.

distinguished from one another in the image, because there is a dependence on the positioning of the slice or pixel, as shown in Figure 9.16. The *partial volume effect* refers to the effect of there being more than one tissue present within a voxel, which is more likely to occur for large voxels and at the edges of objects. The acquired signal will be from a mixture of tissues, and so will be different from that acquired from voxels containing just one tissue or the other. The partial volume effect is illustrated in Figure 9.16c,d.

9.4 Image Noise Characteristics

The signal-to-noise ratio (SNR) in an image is a measure of the relative strengths of the signal and the unwanted noise that is always present in

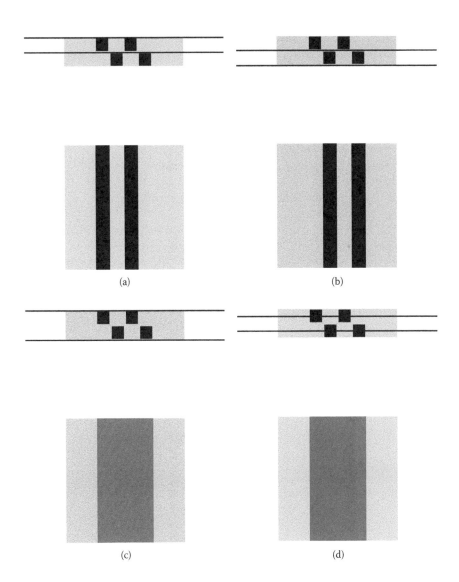

FIGURE 9.16
Effect of slice thickness and position. At the top of each figure a block of tissue containing four objects is shown. The block is the same in each case, and the square cross sections are associated with bars of material that run into the page. The pairs of parallel thin black lines in each figure represent the slice that has been selected for imaging. Underneath is the image that would result. (a and b) A thin slice placed so that just two of the objects are within the selected slice. The objects can be distinguished from one another. (c) A thick slice that includes all four of the adjacent objects. The four objects will not be distinguished in the image. Furthermore, the partial volume effect means that the signal is made up of signals from both the object and the surrounding material and so is different from the signal acquired from the object alone. (d) A thin slice that runs through all four objects. Its position means that once again the objects cannot be distinguished, and the partial volume effect is apparent.

an image. A very low SNR corresponds to having a signal that cannot be distinguished from the background noise in an image, and this is a poor-quality image that will not be useful. Conversely, a high SNR image is desirable. Synthetic examples of SNR were shown in Figure 2.30.

The SNR:

- Increases with the pixel sizes in all three directions
- Increases with the number of measurements made
- Decreases as the receiver bandwidth increases

To make it easier to combine these three factors into a single expression, Equations 9.1 and 9.2 are used.

A voxel is the three-dimensional version of a pixel, so the volume of a voxel is obtained by multiplying together the three pixel sizes:

$$\text{Voxel volume} = V_v = \text{Pixel size}_{SS}\,\text{Pixel size}_{FE}\,\text{Pixel size}_{PE} \qquad (9.1)$$

The number of measurements made is the number of times that the magnetic resonance (MR) signal is sampled to provide information for a particular image. We are already aware that signal acquisition is repeated N_{PE} times and that there are N_{FE} samples taken from each phase encoding step. In addition, it is common to repeat the phase encoding steps so that an average signal can be calculated. The number of repeats is called the number of excitations (NEX), or the number of signals averaged (NSA). Thus,

$$\text{Number of measurements} = N_{meas} = N_{FE}N_{PE}NEX \qquad (9.2)$$

Using these two definitions, the overall expression relating SNR to scan parameters may be written

$$SNR \propto \frac{V_v\sqrt{N_{meas}}}{RBW} \qquad (9.3)$$

Although this expression shows only three parameters, V_v, N_{meas}, and RBW, because of the relationships given in Equations 9.1 and 9.2 there are seven parameters that affect the SNR.

Worked Example 9.3

The two images in Figure 9.17 were acquired using different slice thicknesses, but all the other parameters for the T1-weighted acquisitions were the same. Which image corresponds with a slice thickness of 5 mm, and which with a slice thickness of 1 mm?

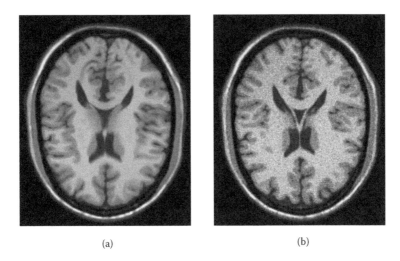

(a)   (b)

FIGURE 9.17
Images with different slice thicknesses for Worked Example 9.3. (Image data from the BrainWeb Simulated Brain Database [1,2]. With permission.)

The image in Figure 9.17a has a higher signal-to-noise ratio (SNR) as it appears smoother and less grainy. SNR is affected by several acquisition parameters, as given in Equation 9.3. In this case, we were told that the only difference between the images was in slice thickness, which is one dimension of voxel volume given by Equation 9.1. SNR is directly proportional to slice thickness, so we expect the SNR to be five times greater in an image with a slice thickness of 5 mm than in an image with a slice thickness of 1 mm. So the image in Figure 9.17a has the greater slice thickness. The poorer through-plane resolution in Figure 9.17 can be seen by considering the triangular region in the center. This region is much better defined in the 1-mm-thick slice in Figure 9.17b, whereas information from the full 5-mm thickness in Figure 9.17a has blurred the triangular shape.

Question 9.10

The two profiles of image intensity shown in Figure 9.18 were both measured across the same row on the images in Figure 9.17. These images were acquired using different slice thicknesses, but all the other parameters for the T1-weighted acquisitions were the same. Which profile is from the image with a slice thickness of 5 mm and which from that with a slice thickness of 1 mm?

Answer

This is essentially the same question as in Worked Example 9.3, but with the information presented in a slightly different way. Profile (a) comes from the image with the greater slice thickness. You can deduce this because this profile appears smoother, and so has a higher signal-to-noise ratio.

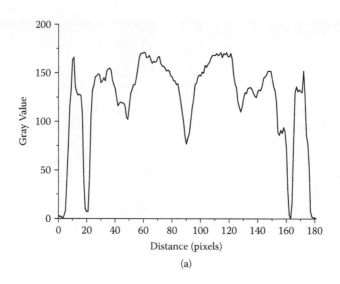

(a)

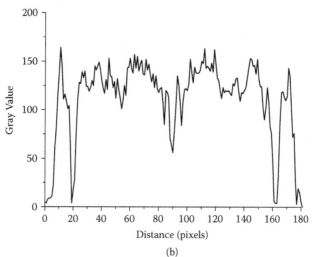

(b)

FIGURE 9.18
Image intensity profiles for Question 9.10.

Question 9.11

The two images in Figure 9.19 represent acquisitions that differ only in the number of excitations (NEX). Which one has been acquired with NEX = 4, and which with NEX = 1?

Answer

In Equation 9.3 we saw that SNR was proportional to the square root of the number of measurements, which is defined in Equation 9.2. In

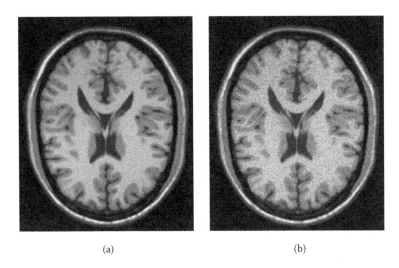

(a)   (b)

FIGURE 9.19
Images for Question 9.11. (Image data from the BrainWeb Simulated Brain Database [1,2]. With permission.)

this case, the only parameter to change its value for the two images is NEX. The image with the larger NEX is expected to have a higher SNR. So the image in Figure 9.19a must have NEX = 4, and the image in Figure 9.19b has NEX = 1.

Question 9.12

What do you expect to happen to spatial resolution and SNR if the FOV is changed to half its previous value, while other parameters, including matrix size, are kept the same?

Answer

If the field of view is halved while maintaining the matrix size, then each in-plane pixel will be half the size, so spatial resolution is increased. From Equation 9.3 the reduction in voxel size will reduce the SNR.

9.5 Scan Time

The total time required to acquire image data using the spin-echo imaging sequence is determined by the values of three of the parameters chosen for the spin-echo sequence according to Equation 9.4:

$$Scan\ time = TR.N_{PE}.NEX \qquad (9.4)$$

TABLE 9.6

Data for Worked Example 9.4

N_{FE}	N_{PE}	NEX	TR/ms	TE/ms	Number of Slices
256	256	1	1,200	15	1
256	256	1	1,200	15	10
256	128	2	500	90	1
128	128	2	500	90	10

TABLE 9.7

Answers for Worked Example 9.4

N_{FE}	N_{PE}	NEX	TR/ms	TE/ms	Number of Slices	Scan Time/ms	Scan Time/min
256	256	1	1,200	15	1	307,200	5.12
256	256	1	1,200	15	10	3,072,000	50.12
256	128	2	500	90	1	128,000	2.13
128	128	2	500	90	10	1,280,000	21.3

Worked Example 9.4

Calculate the scan times for the two-dimensional spin-echo acquisitions in each row of Table 9.6.

For a two-dimensional spin-echo sequence, the acquisition time, or scan time, is given by Equation 9.4. The information provided in Table 9.6 on N_{FE} and TE is therefore not needed. Where only one slice is acquired, Equation 9.4 can be used directly. Where 10 slices are acquired, we will assume that these have been acquired one after the other, so the time required is 10 times that for a single slice. The completed table is Table 9.7.

Question 9.13

The two images in Figure 9.20 were acquired using different matrix sizes. In one the matrix size was 512 × 512, and in the other, 256 × 256. Which is which? Which took longer to acquire?

Answer

The image in Figure 9.20b has the larger matrix size because there have been more pixels acquired over the same field of view. Because of the larger N_{PE}, it would take longer to acquire (Equation 9.4). The image in Figure 9.20b also has a higher spatial resolution. The higher resolution can be seen by looking at the images and arises because there are more pixels in the same distance, meaning that each pixel must be smaller in size.

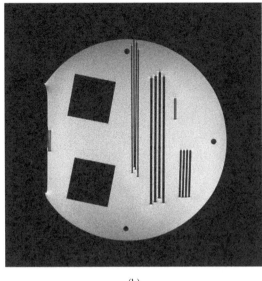

(a)   (b)

FIGURE 9.20
Images for Question 9.13. (Images courtesy of Sarah Bacon.)

Question 9.14

Which of the four T2-weighted images in Figure 9.21 has the small-est value of NEX? Apart from NEX, the acquisitions were made with identical values for the parameters.

Answer

The spatial resolution is known to be the same in each of these images because only NEX differs. As higher NEX values are associated with higher values for the SNR (Equations 9.2 and 9.3), the image with the smallest NEX will be the one with the lowest SNR. The lowest SNR is seen for the image in Figure 9.21b, which therefore is the one with the smallest value of NEX.

Question 9.15

Complete the empty cells in Table 9.8. You may assume that any parameter not listed as a column in the table is the same for all six images, that frequency encoding was in the x direction, and that the FOV is the same in both x and y directions.

Answer

The missing values are shown in Table 9.9. Equation 6.2 was used to find the missing pixel sizes, and N_x for image 5. The SNR for image 4 and NEX for image 6 were found using both Equations 9.2 and 9.3.

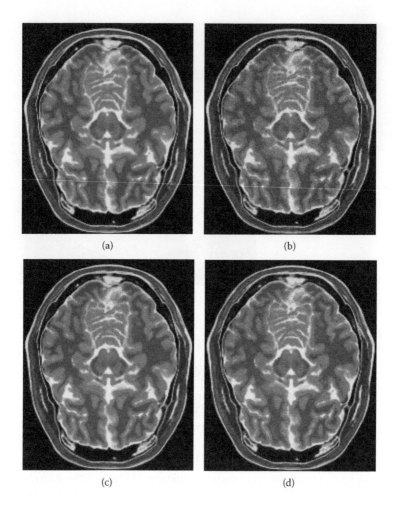

(a)   (b)

(c)   (d)

FIGURE 9.21

Images for Question 9.14. (Image data from the BrainWeb Simulated Brain Database [1,2]. With permission.)

TABLE 9.8

Table for Question 9.15

Image	FOV/mm	N_x	N_y	NEX	SNR	Pixel Size x/mm
1	250	256	256	1	SNR_A	
2	250	512	512	1	$SNR_A/2$	
3	125	256	256	1	$SNR_A/4$	
4	250	256	256	4		0.98
5	250		256	2	$\sqrt{2}\, SNR_A$	1.95
6	250	256	256		$4\, SNR_A$	0.98

TABLE 9.9

Feedback on Question 9.15

Image	FOV/mm	N_x	N_y	NEX	SNR	Pixel Size x/mm
1	250	256	256	1	SNR_A	0.98
2	250	512	512	1	$SNR_A/2$	0.49
3	125	256	256	1	$SNR_A/4$	0.49
4	250	256	256	4	$2\,SNR_A$	0.98
5	250	**128**	256	2	$\sqrt{2}\,SNR_A$	1.95
6	250	256	256	16	$4\,SNR_A$	0.98

9.6 The Spin-Echo Image Simulator

9.6.1 Weighting for Brain Imaging

In this simulation a synthetic brain image is calculated showing the relative gray values expected for gray matter, white matter, fat, and CSF if a spin-echo sequence is used for imaging. You can select TE and TR using the sliders, and a new image is automatically constructed.

Start the Image Simulator–Spin-Echo program as explained in Chapter 1. Leave the *Image Selection* drop-down menu as it is, showing *brain1*. To get used to using the interactive simulator, try setting TR and TE to some of the combinations suggested in Section 9.2.1 for acquiring proton density-weighted, T1-weighted, and T2-weighted images. Examples of suitable values are given in Table 9.10. If you type the value in, remember to finish the entry using the enter key. From your knowledge of the appearance of the differently weighted images, and bearing in mind that the demonstration represents a simplified version of the tissue distribution, do your results look correct? If the images look dark or lack contrast, they may be windowed using the sliders labeled *Window/Level Adjuster* (see "Window and Level" box).

WINDOW AND LEVEL

The principles of windowing an image display were introduced in Section 2.4.2.1. Where an image lacks contrast between gray levels, make the display window narrower by reducing the window width, using the upper of the two sliders in the simulator. Then adjust the gray level on which the window is centered by changing the window level. Repeat until the image appears as desired. For example, when TR = 550 ms and TE = 14 ms, a window of width 70 centered on a level of 35 works well. In the simulator the window is reset each time the acquisition parameters are changed.

TABLE 9.10

Suitable Values of TR and TE to Use in
the Spin-Echo Image Simulator

	TR/ms	TE/ms
T1-weighting	550	14
T2-weighting	2,500	90
Proton density weighting	2,500	15

Real images acquired using the parameters in Table 9.10 are compared with those generated by the simulator program in Figure 9.22. The simulator gives reasonable results for T1-weighted and T2-weighted images, where image contrast is dominated by the relaxation times. However, it does not work well for proton density weighting, because in the simulator all the tissues have the same proton density.

The values of T1 and T2 that are set in the simulator for the four tissues are shown in Table 9.11. Proton density is assumed to be the same for all of the tissues. The tissues are shown labeled in Figure 9.23.

Question 9.16

Use the information in Table 9.11 to draw sketch graphs (as outlined in Section 9.2.2) and decide whether TE and TR should be long or short to make CSF brighter than gray matter. Check your prediction using the simulator. Is this a T1-weighted or T2-weighted image?

Answer

In Table 9.11 it can be seen that both T1 and T2 are longer for CSF than for gray matter, so you can sketch recovery curves for materials with long and short relaxation times, as in Figures 9.2, 9.3, and 9.4. In both the short and long TR sketches, CSF is brighter than gray matter if a long TE is used. The CSF signal will be larger if a long TR is used, as more recovery will have taken place. So, both TR and TE should be long. The longest values available in the simulator are TR = 3,000 ms and TE = 100 ms, but other, slightly different values are quite acceptable. The image given by the simulator for these values is shown in

TABLE 9.11

The T1 and T2 Values for Four Types of Brain
Tissue in the Spin-Echo Image Simulator

Material	T1/ms	T2/ms	Proton Density
Fat	260	84	100%
White matter	780	90	100%
Gray matter	920	100	100%
CSF	2,000	300	100%

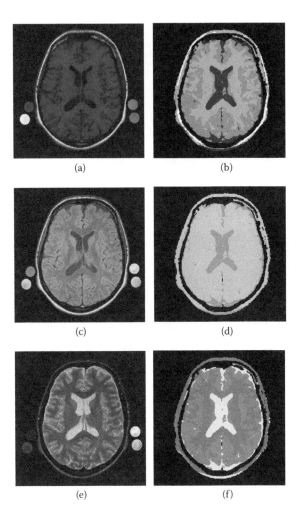

(a) (b)

(c) (d)

(e) (f)

FIGURE 9.22
Comparison of real spin-echo brain images and those generated in the spin-echo image simulator. (a) Real T1-weighted image with TE = 550 ms and TE = 14 ms. (b) Simulator T1-weighted image with TE = 550 ms and TE = 14 ms. (c) Real proton density-weighted image with TE = 2,500 ms and TE = 15 ms. (d) Simulator proton density-weighted image with TE = 2,500 ms and TE = 15 ms. (e) Real T2-weighted image with TE = 2,500 ms and TE = 90 ms. (f) Simulator T2-weighted image with TE = 2,500 ms and TE = 90 ms. All the images have been windowed for display. The images that differ most between real acquisition and simulation are those from proton density weighting; as in the simulator, all the tissues have the same proton density. The four circular features in the real images are tubes with known values of T1 and T2. Tubes of this kind may be used for quality assurance purposes.

Figure 9.24a. This is a T2-weighted image. Materials with long T2 are always bright in such images, and the effect of T1 has been reduced by choosing a long TR, which means that full recovery occurs for materials with both long and short T1 values.

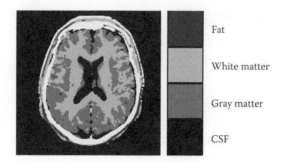

FIGURE 9.23
The simplified distribution of tissues in the simulator *brain1* image.

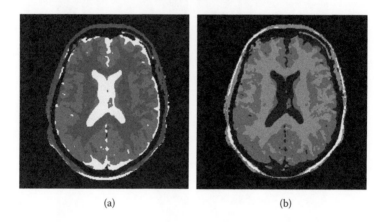

(a)   (b)

FIGURE 9.24
(a) Simulator-generated T2-weighted brain image with TE = 3,000 ms and TE = 100 ms. (b) Simulator-generated T1-weighted brain image with TE = 500 ms and TE = 10 ms. Both images were displayed with a window of width 60 and level 90.

Question 9.17

Use the information in Table 9.11 to draw sketch graphs (as outlined in Section 9.2.2) and decide whether TE and TR should be long or short to make gray matter brighter than CSF. Check your prediction using the simulator. Is this a T1- or T2-weighted image?

Answer

It can be seen from the sketch graphs that only a short TR with short TE will give brighter gray matter than CSF. Values TR = 500 ms and TE = 10 ms give the image shown in Figure 9.24b. This is a T1-weighted image. Materials with short T1 are always bright in such images, and the effect of T2 has been reduced by choosing a short TE, which means that the different signals arising from the differing T1 values are measured.

You can explore the crossover effect seen in the short TR sketch (Figure 9.4) by increasing the value of TE. Set TR to 1,500 ms, and gradually increase TE from 10 to 100 ms. Initially, gray matter will be brighter than CSF, but there will be a change to CSF being brighter than gray matter for longer TE. You should see the contrast between CSF and gray matter flip when TE is about 62 ms. We will explore the crossover point itself, where the signal from two tissues is the same, in the following exercises with the image simulator.

9.6.2 Acquisition at the Crossover Point

On the *Image Selection* drop-down menu select *brain2* and set TR to 600 ms and TE to 30 ms. From the values of TR and TE, do you expect this to be a T1-weighted or a T2-weighted image?

Both times are fairly short, so a T1-weighted image is expected. The bright signal from the fat behind the eyes on the simulations confirms this.

Leave TR and TE as they are and on the *Image Selection* drop-down menu select *face*. You should see a sleeping face with closed eyes as shown in Figure 9.25a. Do the eyelashes have a longer or shorter T1 value than the surrounding face?

Because we know that this is a T1-weighted image, we know that T2 has only a small influence on the brightness seen in the image. The brighter signal from the lashes suggests that their T1 is shorter than the T1 of the surrounding face.

On the *Image Selection* drop-down menu select *brain1* and set TR to 2,500 ms and TE to 70 ms.

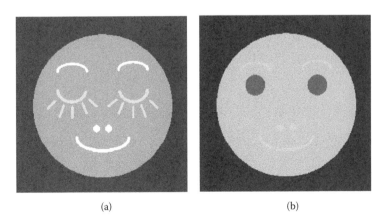

(a)   (b)

FIGURE 9.25
The simulator *face* image: (a) T1-weighted, TR = 600 ms, TE = 30 ms; (b) T2-weighted, TR = 2,500 ms, TE = 70 ms. Both images were displayed with a window of width 120 and level 40.

From the values of TR and TE, do you expect this to be a T1-weighted or T2-weighted image?

Both times are long, so a T2-weighted image is expected. The bright signal from the CSF on the simulations confirms this.

Leave TR and TE as they are and on the *Image Selection* drop-down menu select *face*. You should see a face with open eyes as shown in Figure 9.25b. Do the eyes have a longer or shorter T2 value than the surrounding face?

We know that this is a T2-weighted image, so we know that T1 has only a small influence on the brightness seen in the image. The brighter signal from the face suggests that the face has a longer T2 than the eyes.

The eyelashes, which we saw in the T1-weighted image, are not visible in this T2-weighted image. This is because the image was acquired at a time when the signal from the face and the signal from the eyelashes were identical. Similarly, in the T1-weighted image, the eyes were not seen because the signals from both the face and the eyes were the same for the selected values of TR and TE. In each case, acquisition was made at the crossover point of the face curve and the curve for one of the other materials.

Return to the T1-weighted settings (TR = 600 ms and TE = 30 ms). Move the TE slider from side to side and confirm that the eyes reappear. Are the eyes brighter or darker than the face when TE < 30 ms is used? Use the earlier deductions about the relative sizes of T1 and T2, and assuming that the lashes and eyes have the same T1, draw sketch graphs to confirm that this is what is expected.

We saw earlier that T1 for the face was longer than T1 for the eyelashes, and because the eyelashes and the eyes have the same T1, we can conclude that T1 for the face is also longer than T1 for the eyes. We have also previously deduced that T2 for the face is longer than T2 for the eyes. Sketch 3, for a T1-weighted, short TR image, and based on this knowledge, is in Figure 9.26. The sketch confirms that at shorter TE, the eyes will be brighter than the face.

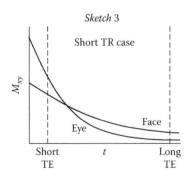

FIGURE 9.26

Sketch 3 drawn for the eye and face components of the synthetic image that is used to demonstrate acquisition at the crossover point.

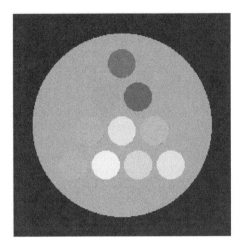

FIGURE 9.27
The spin-echo image simulator showing the test object. TR = 2,500 ms, TE = 50 ms. A window of width 150 and level 75 was used.

9.6.3 Weighting in a Test Object

On the *Image Selection* drop-down menu select *test object* and set TR to 2,500 ms and TE to 50 ms. The image is the simulation of a slice through a test object that contains 10 cylindrical tubes arranged in a larger cylinder (Figure 9.27). The tubes can be identified using the numbering shown in Figure 9.28.

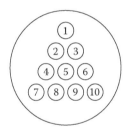

FIGURE 9.28
The numbering of the tubes in the simulator test object.

Question 9.18

Use the simulator and your knowledge of T1-weighting and T2-weighting to answer the following questions:

(a) Tubes 2 and 3 have the same T1 value. Which tube has the longer T2?
(b) Tubes 4 and 7 have the same T2 value. Which tube has the longer T1?
(c) Do tubes 8 and 9 have the same T1 value or the same T2 value?
(d) Which tube has the shortest T1 value and which the longest T2 value?

Answer

(a) In a T2-weighted image, long T2 materials appear brighter. If the simulator is set up to give a T2-weighted image (long TR, long TE), tube 2 appears brighter than tube 3, and therefore has the longer T2 value.

(b) In a T1-weighted image, short T1 materials appear brighter. If the simulator is set up to give a T1-weighted image (short TR, short TE), tube 4 appears brighter than tube 7, and therefore has the shorter T1 value. T1 is longer for tube 7.

(c) The brightness of tubes 8 and 9 is similar in a T1-weighted image, but differs in a T2-weighted image. The two tubes share the same value of T1.

(d) The tube that appears brightest in a T1-weighted image has the shortest T1; this is tube 1. The tube that appears brightest in a T2-weighted image has the longest T2; this is tube 8.

This simulated test object has similar characteristics to the TO5 test object [4] that may be used to check T1 and T2 image contrast in real systems. Other measures of imaging performance are outlined in the next section.

TO CLOSE THE PROGRAM

Click on the cross at the top right of the window, and then confirm that you want to end the program.

9.7 System Performance Assessment

It has been seen in this chapter that the quality of an MR image is dependent on a large number of parameters associated with the acquisitions, and there are many other system-related factors involved too. As a result, quality assurance has an important part in clinical practice. System performance is assessed when equipment is first installed, and periodically after that to detect any deterioration in performance. Measurement protocols have been specified by a range of organizations [5]. The protocols use test objects to assess parameters, including signal uniformity, SNR, geometrical properties, and image contrast. Each test object is specially designed for the task and has known properties. The test object in Figure 9.20 is an example that may be used for assessing spatial resolution.

9.8 Chapter Summary

The focus of this chapter was the factors that influence the appearance of an image acquired using the spin-echo imaging sequence. The important topic of image weighting, which determines the gray-scale appearance of the image, was considered. The link between image weighting and saturation recovery plots was emphasized, and a range of worked examples, questions, and interactive exercises using the Spin-Echo Image Simulator were employed to promote understanding. The appearance of the image in terms of spatial resolution and noise is affected by a number of parameters, and these effects were described both qualitatively and using equations. In the next chapter the familiarity with the ideas behind T1- and T2-weighting that was acquired in this chapter will be applied to two different imaging sequences.

References

1. The BrainWeb Simulated Brain Database. www.bic.mcgill.ca/brainweb/.
2. Collins, D. L., et al. 1998. Design and construction of a realistic digital brain phantom. *IEEE Trans. Med. Imag.* 17:463.
3. Westbrook, C. and Kaut, C. 1993. *MRI in practice*, chap. 2. Oxford: Blackwell Science.
4. Lerski, R. A. and de Certaines, J. D., II. 1993. Performance assessment and quality control in MRI by Eurospin test objects and protocols. *Magn. Reson. Imag.* 11:817.
5. McRobbie, D. W., et al. 2003. *MRI: From picture to proton*, chap. 11. Cambridge: Cambridge University Press.

10

Further Imaging Sequences

In this chapter no further new concepts are introduced. Instead, material from the earlier chapters is brought together to introduce two further imaging sequences and to describe the effect of flowing blood on images. The inversion recovery imaging sequence, and its use to null the signal from a selected tissue, is covered first. Two programs associated with inversion recovery are used here: the *Inversion Recovery* program, which plots recovery curves, and the *Image Simulator–Inversion Recovery* program. The second new imaging sequence is the gradient-echo imaging sequence, which is faster than spin-echo imaging. The chapter continues with a discussion of flow, covering both its effects on an image and the techniques used to image flowing blood. The interactive *Flow Phenomena* program is used to support the description of flow phenomena in the spin-echo sequence, especially concerning the relationships among the flow velocity, slice thickness, and time to echo (TE).

10.1 Learning Outcomes

When you have worked through this chapter, you should:

- Be familiar with the principles of two more imaging pulse sequences
- Understand how blood flow across a slice affects an MR image

10.2 Inversion Recovery Sequence

10.2.1 Principles of Inversion Recovery

The inversion recovery sequence is an adaptation of the saturation recovery sequence. The adaptation is made so that the operator can arrange for there to be no signal from a particular tissue. Setting the signal to zero in this way is called signal *nulling*. Although it may seem odd to deliberately avoid getting a signal, it can be helpful in practice. A good example is the removal of the strong signal that arises from fat, so that subtle differences between other tissues can be seen.

In Chapter 4 we saw that saturation can be achieved by applying a 90° radio frequency (RF) pulse to tip the net magnetization vector into the $x'y'$ plane, leaving no net magnetization in the z direction. Saturation recovery is the process of recovery from the state of saturation, and M_z increases from zero. In a similar way, *inversion recovery* is the process of recovery from the state of *inversion*. Inversion is the result of applying a 180° RF pulse, so for inversion recovery M_z increases from a maximum negative value, rather than from zero, which is the case for saturation recovery. It can be seen in Figure 10.1 that following the 180° RF pulse, M_z has the value zero at a time that depends on the value of T1 for the material.

Just like in the saturation recovery sequence, the next step in the inversion recovery pulse sequence is a 90° RF pulse to tip M_z into the $x'y'$ plane. The time between the initial 180° RF pulse and the 90° RF pulse is called the *time to inversion* (TI) or *inversion time*. Although the symbols can look very similar, do not confuse this pulse sequence parameter with T1, the longitudinal relaxation time, which is a property of a material. As in the saturation recovery pulse sequence, measurement is made at a time TE after the 90° RF pulse (Figure 10.2). The time between successive 180° RF pulses is TR (the repetition time).

In the saturation recovery sequence all materials, whatever their T1 value, have $M_z = 0$ at the time of the first RF pulse. The main difference from the saturation recovery sequence is that in inversion recovery, each material has $M_z = 0$ at a characteristic time after the inverting 180° RF pulse. This time is known as the *null point*. If the inversion time (TI) is set

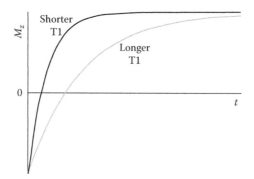

FIGURE 10.1
Graph showing the z component of the net magnetization vector recovering after a 180° RF pulse, for two materials with different T1 values.

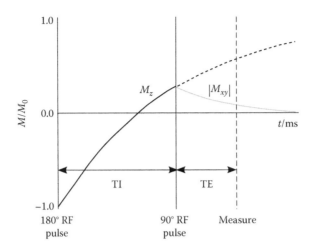

FIGURE 10.2
An example inversion recovery graph to illustrate the key features of graphs of this kind. The plot shows M_z to the left of the solid vertical line, and $|M_{xy}|$ to the right of it. In this sketch they have been differentiated by using black for M_z and gray for $|M_{xy}|$. A 180° RF pulse is applied at time zero, and a 90° RF pulse corresponds with the solid vertical line at a time TI later. Measurement is made at a time TE after the 90° RF pulse.

so the 90° RF pulse is applied when $M_z = 0$ for a particular material, then no signal will be tipped into the $x'y'$ plane, and no signal will be measured for that material. The signal has been *nulled* or *suppressed*.

10.2.2 Magnitude Signals

Both positive and negative values of M_z are present when the inversion recovery sequence is used. In the diagrams in this chapter we adopt the convention of showing the absolute value of M_{xy} after the inversion pulse (Figure 10.3). The absolute value (magnitude) is indicated using vertical bars, $|M_{xy}|$, and is the value of M_{xy}, ignoring its sign.

10.2.3 The Inversion Recovery Program

The Inversion Recovery program is a lot like the Saturation Recovery program that you first used in Chapter 4. Start the Inversion Recovery program as indicated in Chapter 1. The display should look like Figure 10.4. Click on *Tissue A (black)* in the middle column and a graph appears in the lower half of the interface (Figure 10.5). The plot returned by the inversion recovery program illustrates relaxation starting immediately following a 180° RF pulse:

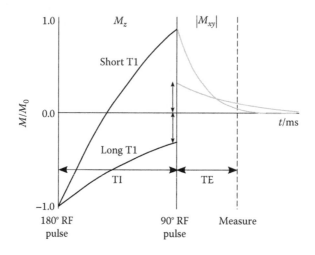

FIGURE 10.3

The material with the longer T1 still has a negative value of M_z at the time of the 90° RF pulse. When the absolute or magnitude value is used, the sign of M_z before it is flipped into the xy plane is ignored, so that all $|M_{xy}|$ values are positive. In this example, the vertical arrows indicate the magnitude of M_z for the longer T1 material.

- At time $t = 0$, a 180° RF pulse is applied to tip the net magnetization vector from alignment with the positive z axis to alignment with the negative z axis, that is, to invert the magnetization.

- In the first part of the plot, between $t = 0$ and the first vertical black line, the vertical axis indicates the value of M_z, which is the z component of the net magnetization vector. In this case, the vertical black line is at time $t = 600$ ms.

- At $t = $ TI (the first vertical black line), a 90° RF pulse is applied to tip the magnetization vector into the $x'y'$ plane. Note how the amount of magnetization available to this second pulse is dependent on the amount of relaxation after the first pulse. The time period is called the inversion time (TI).

- In the part of the plot after the first vertical black line, the vertical axis indicates $|M_{xy}|$ (instead of M_z, which is plotted before the first vertical black line). As we have seen in preceding chapters, M_{xy} is the component of interest because the magnetic resonance (MR) signal is measured in the $x'y'$ plane.

- The value of $|M_{xy}|$ at $t = $ TI is the same as the absolute value of M_z immediately before the 90° RF pulse, because a 90° RF pulse tips all the available z magnetization into the $x'y'$ plane.

- The component $|M_{xy}|$ then decays. The signal is measured at the time of the second (rightmost) vertical black line.

FIGURE 10.4
The graphical user interface for the Inversion Recovery program.

- For consistency with saturation recovery and spin-echo sequences, the time period between the 90° RF pulse and measurement is called TE. The time between successive 180° RF pulses is TR.

- As with the Saturation Recovery program, it is possible to set a long TE in the program so that the second black line appears far to the right, and it is then possible to use the plot to consider the effects of using both short and longer TEs.

- The curve in Figure 10.5 is for TR > 3,000 ms (TR is fixed in the program), TI = 600 ms, and TE = 50 ms and a material that has T1 = 722 ms and T2 = 100 ms.

- The dotted line that continues the first curve is there to emphasize the course of T1 recovery that would take place if the 90° RF pulse were not applied.

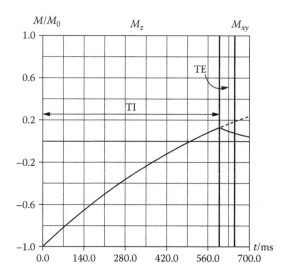

FIGURE 10.5
An inversion recovery plot generated by the Inversion Recovery program for a material with
T1 = 722 ms and T2 = 100 ms. The time periods TI and TE are indicated in the figure but
do not appear when the program is run. TR > 3,000 ms (this is fixed in the program), TI =
600 ms, and TE = 50 ms.

Question 10.1

(a) Is the parameter shown in the first part of the plot, from $t = 0$ to
$t =$ TI, M_z, M_{xy}, $|M_z|$, or $|M_{xy}|$?
(b) Is the parameter shown in the second part of the plot, for $t >$ TI,
M_z, M_{xy}, $|M_z|$, or $|M_{xy}|$?
(c) Is the component of magnetization that gives rise to a signal
that can be detected in an MR system M_z or M_{xy}?

Answer

The statements in Question 10.1 should have reminded you about the
key features of the graphs plotted using the Inversion Recovery program. The parameter shown in the first part of the plot, from $t = 0$ to
$t =$ TI, is M_z. The parameter shown in the second part of the plot, for
$t >$ TI, is $|M_{xy}|$. The component of magnetization that gives rise to a
signal that can be detected in an MR system is M_{xy}.

Now, use the program to generate some new curves. Put ticks in the
boxes in the *Choose the tissue/s to display* area of the interface for fat, cerebrospinal fluid (CSF), gray matter, and white matter. Remove the tick for tissue A. Leave the other settings as they are. Curves will appear for the
selected tissues (Figure 10.6). The curves appear in the colors indicated
beside the tick boxes.

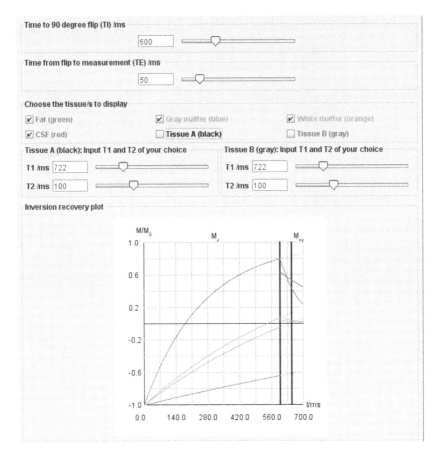

FIGURE 10.6
The graphical user interface for the Inversion Recovery program when the four named tissues are selected.

Question 10.2

Use the plot that you have just generated to answer the following questions:

(a) Which has the longer T1 relaxation time, fat or gray matter?
(b) Which has the longer T1 relaxation time, white matter or gray matter?
(c) If an image were acquired at the time indicated by the second vertical black line, which of the four tissues would appear brightest in the image?

Answer

The first section of the plot is the part to concentrate on for parts (a) and (b) of the question because it shows M_z, and thus indicates T1 (longitudinal or spin-lattice) relaxation. The slope of the gray matter curve in the first section of the plot is less steep than the slope of the fat curve, indicating the longer T1 relaxation time for gray matter. We can deduce that the T1 relaxation times for white matter and gray matter must be very similar, as the orange and blue curves have a similar slope in the first section of the plot. The blue curve is slightly below the orange one, which suggests a slightly longer T1 relaxation time for gray matter. Look at the second section of the plot, which shows $|M_{xy}|$ and indicates T2 (transverse or spin-spin) relaxation, to answer part (c). M_{xy} is the component that is measured as the MR signal, so the highest value leads to the brightest pixels in an image. In this case, at the measurement time indicated by the vertical black line, CSF (red) would appear the brightest of the four tissues.

The values for T1 and T2 that have been set in the Inversion Recovery program are the same as those in the saturation recovery program, and are shown in Table 4.2. The values shown in the table confirm that T1 is longer for gray matter than for fat, and longer for gray matter than for white matter.

Question 10.3

Adjust TI and TE in the inversion recovery program so that the signal from the white matter is nulled.

Answer

The concept of nulling, or suppressing, a signal was outlined in Section 10.2.1. To null the signal from white matter, the method is to apply the 90° RF pulse when the magnetization vector for white matter has recovered through 90° and has an M_z component of zero. If a 90° RF pulse is applied at this time to tip the M_z component into the $x'y'$ plane for measurement, the resulting signal is zero. Thus, the signal from the white matter has been nulled. In the program, it can be seen that the point on the white matter recovery curve where $M_z = 0$ is at about $t = 545$ ms, so try setting TI to 545 ms. With TI set to this value, the signal from the white matter will be zero whatever value of TE is chosen.

The most common use of signal nulling is to remove the strong signal from fat. Set TI to 180 ms in order to null fat (Figure 10.7), and this time, the green fat line is zero for all values of TE. In this example the other curves are not close to the one for fat, but in practice it is usual to null fat if the fat signal is making it hard to assess an adjacent feature of interest because they have similar values. Examples include tumor in bone marrow or in the orbit of the eye.

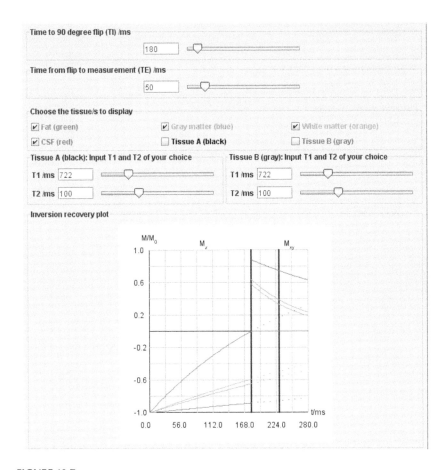

FIGURE 10.7
The graphical user interface for the Inversion Recovery program set up to null the signal from fat.

Select the tick box for tissue A, and where it says *Tissue A (black): Input T1 and T2 of your choice,* set T1 to 300 ms and T2 to 100 ms. These are values similar to, but not identical to, those for fat. The relevant curve will be plotted in black. First set TI to 500 ms so that the fat signal is not nulled, and leave TE at 50 ms. In your result, it can be seen that at the time of measurement the black curve is very close to the green fat curve, and it will be hard to distinguish the signal from the two materials (Figure 10.8). Now set TI to 180 ms to null the fat signal. With these settings, although the black and green curves are close together, they are better separated than before, and at the measurement time there is a clear difference between them, because of the nulling of the fat signal.

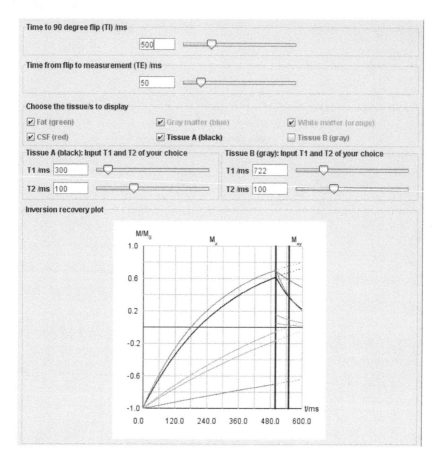

FIGURE 10.8
The graphical user interface for the Inversion Recovery program showing a material with relaxation times similar to those of fat also present.

VALUE OF TI TO NULL THE SIGNAL

The recovery of M_z from inversion is described by the expression

$$M_z = \left(1 - 2\exp\left(-\frac{t}{T1}\right)\right)$$

To null the signal, $M_z = 0$ when $t = $ TI. By substitution it can be shown that $TI = 0.693T1$; that is, to null the signal from a tissue, TI should be set to a value approximately 0.693 times the T1 of the tissue to be nulled.

10.2.4 The Inversion Recovery Imaging Pulse Sequence

The inversion recovery imaging pulse sequence is shown in Figure 10.9.
The sequence uses the same method as the spin-echo sequence to generate an echo, and may be thought of as a spin-echo sequence where there
is an additional inverting 180° RF pulse, included at a time TI before the
initial 90° RF pulse. The timing diagram is laid out in the same way as the
one in Figure 8.10 for the spin-echo sequence.

- In the top row:
  - The inverting 180° RF pulse is shown, repeated at an interval
    of TR.
  - The 90° RF pulse that tips the magnetization into the xy plane
    occurs at a time TI after the inverting pulse.
  - Then, as in the spin-echo sequence, a rephasing 180° RF pulse
    is applied at a time TE/2 after the 90° RF pulse.
- The next row shows the G_{ss}, or slice select, gradient. It is applied at
  the time of the 90° RF pulse so that only the spins in that selected
  slice are saturated and can contribute to the signal.

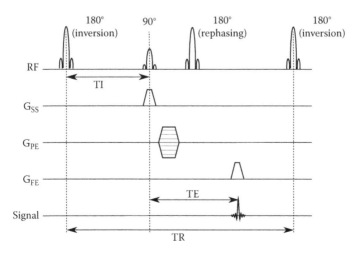

FIGURE 10.9
Simplified timing diagram for the inversion recovery imaging pulse sequence.

- The phase encoding gradient is applied once per repetition, with a different strength on each occasion.
- The frequency encoding, or readout, gradient is timed to coincide with the time of the echo.
- The last row shows when the signal is measured, at a time TE after the 90° RF pulse.

Like Figure 8.10 for the spin-echo sequence, the diagram of Figure 10.9 is simplified, because it shows the gradients being switched on and off in a way that will permit spatial localization, and no others. It does not show the additional gradients that are applied in practice to compensate for the dephasing effects of the gradients themselves, and for the effects of the RF pulses.

Differently weighted images can be achieved using the inversion recovery sequence (Table 10.1). Unlike the spin-echo sequence, a short TR is not used to achieve T1-weighting emphasizing differences in T1. In the inversion recovery sequence, long TRs are always used to ensure that there is full recovery between inverting pulses; otherwise, the images become hard to interpret. T1-weighting relies on using a short TE, to minimize the influence of T2, and midlength TI, to emphasize the influence of T1 (Figure 10.10a). To see T2 differences in an image that is still predominantly T1-weighted, a large value for TE is used with a midlength TI (Figure 10.10b). Proton density weighting is achieved using a long TI and a short TE, to reduce the impacts of both T1 and T2 (Figure 10.10c).

The variation of imaging sequence that incorporates the ideas of both inversion and signal nulling is the short TI inversion recovery (STIR) sequence. As implied by the name, the inversion time (TI) has a small value that is suitable for nulling the signal from fat. STIR images are usually T1-weighted because of the short TI, but as for the standard inversion recovery sequence, it is possible to see the effects of T2 by increasing TE, and this can be helpful for emphasizing pathology in the images. The T in the STIR acronym is sometimes said to represent τ (tau), where τ represents inversion time.

TABLE 10.1

Timing Parameters and Weighting in the Inversion Recovery Imaging Sequence

	TR	TE	TI
T1-weighted image	Long	Short	Mid
T2-weighted image	Long	Long	Mid
Proton density-weighted image	Long	Short	Short

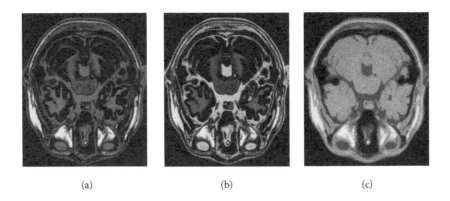

(a)   (b)   (c)

FIGURE 10.10
Examples of images generated using the inversion recovery imaging pulse sequence. (a) T1-weighted image with TR = 2,000 ms, TI = 500 ms, and TE = 15 ms. (b) Pathology-weighted image with TR = 2,000 ms, TI = 500 ms, and TE = 80 ms. (c) Proton density-weighted image with TR = 2,000 ms, TI = 1,500 ms, and TE = 15 ms. (Image data from the BrainWeb Simulated Brain Database [1,2]. With permission.)

Question 10.4

Figure 10.11a is a T1-weighted spin-echo image in which fat appears bright. The two images in Figure 10.11b and c were obtained using inversion recovery sequences. In both cases, TR = 2,000 ms and TE = 15 ms. Which of the two images has TI = 150 ms, and which has TI = 500 ms?

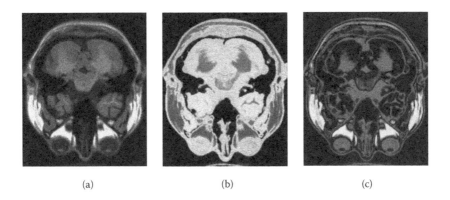

(a)   (b)   (c)

FIGURE 10.11
Images for Question 10.4. (Image data from the BrainWeb Simulated Brain Database [1,2]. With permission.)

Answer

The noticeable feature on the spin-echo T1-weighted image (Figure 10.11a) is the bright fat signal behind the eyes. Inversion recovery images are also usually T1-weighted, but in addition, TI can be chosen to null the signal from a material with a particular T1 value. Figure 10.11c has bright fat just as in the spin-echo image, but in Figure 10.11b the signal from fat is very much reduced, suggesting that this image is the one for which TI = 150 ms (approximately 0.693 times the value of T1 for fat). Recovery curves generated using the inversion recovery program and corresponding to the two images are shown in Figure 10.12. In each case the fat curve is shown. Note how much smaller the signal from fat is at the time of measurement in the TI = 150 ms sequence (Figure 10.12a). Run the inversion recovery program for yourself with the other tissues selected too. You will see that when a short TI is used rather than a long TI, the other curves, which represent other components of the brain, have larger signals and differ less from one another.

Question 10.5

The curves shown in Figure 10.13a and b are both for inversion recovery sequences with TR = 3,000 ms and TI = 500 ms. The two sequences have different values for TE. TE is longer in Figure 10.13a.

(a) Would you expect the signal from fat to be nulled in either of these sequences?
(b) Which sequence do you expect to show predominantly T1-weighting, with very little influence from T2?
(c) Why might you choose to use the longer TE as is done in Figure 10.13a?

Answer

(a) You would not expect the signal from fat to be nulled by either of these sequences, which have TI = 500 ms. To null fat, a short TI of about 150 to 180 ms is required (0.693 times T1 for fat).
(b) The sequence with the shorter TE should show predominantly T1-weighting, as differences from differing T2 decay times will not have time to manifest themselves. This is the case in Figure 10.13b.
(c) Allowing some T2-weighting on a predominantly T1-weighted inversion recovery image, by using a longer TE, is a way of showing up diseased tissue in which T2 is lengthened compared with surrounding tissue, perhaps because of increased water content.

Question 10.6

The contrast agent gadolinium can result in other tissues having a T1 value similar to that for fat. Is the STIR approach to nulling the signal for fat a useful one in a contrast-enhanced image?

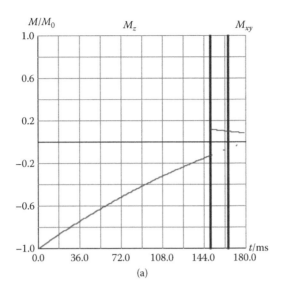

(a)

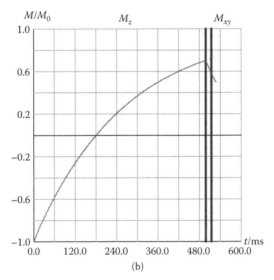

(b)

FIGURE 10.12

Recovery curves generated using the inversion recovery program corresponding to the images in Figure 10.11b and c. In each case the curve shown is the fat curve. (a) TI = 150 ms, (b) TI = 500 ms.

Answer

It would be sensible to avoid using STIR on a contrast-enhanced image, because the contrast agent may have affected the T1 of the tissue of interest so that it has a value that will cause its signal to be nulled.

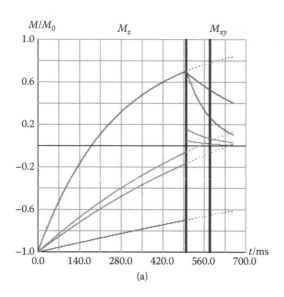

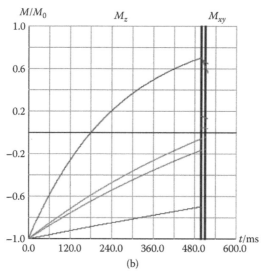

FIGURE 10.13
Recovery curves for Question 10.5, generated using the inversion recovery program.

10.3 The Inversion Recovery Image Simulator

10.3.1 Weighting for Brain Imaging

The Inversion Recovery Image Simulator works in the same way as the Spin-Echo Image Simulator introduced in Chapter 9. A synthetic brain image is calculated showing the relative gray values expected for gray matter, white matter, fat, and cerebrospinal fluid (CSF) if an inversion recovery imaging sequence, like that shown in Figure 10.9, is used for imaging. In the program, TR, TE, and TI may be set using typed input or sliders, and a new image is automatically constructed. The same values of T1 and T2 are used for the tissues as for the Spin-Echo Simulator of Chapter 9 (Table 9.11). The tissues are shown labeled in Figure 9.23, and proton density is assumed to be the same for all the tissues.

Start the Image Simulator–Inversion Recovery program as explained in Chapter 1. Leave the *Image Selection* drop-down menu as it is, showing *brain1*. Set TR, TE, and TI first to the combinations used in Figure 10.10 for T1-weighted images, and then to the values for pathology-weighted images. Take care to enter the values in the correct order. You will need to adjust the level and window on the demonstrator to improve the appearance of the displayed images. The simulator does not include any differences in proton density, so it is not suggested that you try to reproduce proton density weighting.

10.3.2 Nulling Fat in Brain Imaging

In the *Image Selection* drop-down menu, select *brain2*. This is an image at the level of the eyes where suborbital fat is easily seen. Set TR to 3,000 ms, TE to 50 ms, and TI to 180 ms. These settings correspond with those used to generate the recovery curves in Figure 10.7—the fat is nulled and cannot be seen on the image. Adjust the TI value slowly using the slider, and you should see the fat signal reappear for higher and lower values of TI.

10.3.3 Nulling and Pathology Weighting in a Test Object

It was noted that STIR images are strongly T1-weighted because of the short inversion times used, but that a longer TE is sometimes chosen to show diseased tissue that has a longer T2 than surrounding, similar tissue. In the *Image Selection* drop-down menu, select *test object*. Set TR to 3,000 ms, TE to 20 ms, and TI to 300 ms (Figure 10.14a). In this exercise we shall concentrate on tubes 1, 2, and 3 (Figure 9.28). Imagine that these are tissues in a clinical image. All three are giving a visible signal, but we know:

- The tissue represented by tube 1 is healthy and is different from the tissue represented by tubes 2 and 3.
- Tubes 2 and 3 represent the same tissue, and tube 2 represents diseased parts of this tissue characterized by an increased T2.

The first step is to null the signal from tube 1, the healthy tissue that may be hiding an interesting signal. T1 for tube 1 is 150 ms, so set TI in

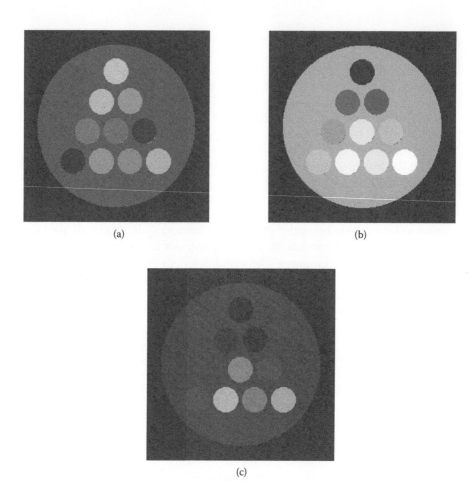

FIGURE 10.14

Simulator-generated inversion recovery images of a test object. (a) TR = 3,000 ms, TE = 20 ms, TI = 300 ms. (b) TR = 3,000 ms, TE = 20 ms, TI = 104 ms to demonstrate nulling of the signal in tube 1. (c) TR = 3,000 ms, TE = 150 ms, TI = 104 ms to demonstrate weighting together with nulling. There is a slightly greater signal from tube 2 than from tube 3. All the images used a window of width 200 and level 50. The difference between tubes 2 and 3 in (c) can be emphasized in the simulator by using width 20 and level 10.

the simulator to 104 ms. The signal will be reduced from all the tubes as a result, but the windowing can be adjusted by reducing the window width and level. In the result (Figure 10.14b), we see that the signals from tubes 2 and 3 are still similar, but in the clinical image these would no longer be hidden by the signal from the tissue corresponding with tube 1, which is now nulled.

To include some T2 influence in the image, TE is increased. Set TE to 150 ms (Figure 10.14c). The signal from tube 1 remains nulled, while the signal from tube 3, although now very small, is better distinguished from tube 2. Tube 2 represents diseased tissue with an increased T2, and it is now slightly brighter than tube 3.

10.4 The Gradient-Echo Sequence

10.4.1 Principles of the Gradient-Echo Sequence

The spin-echo sequence has drawbacks for clinical imaging, the main issue being that it is a slow acquisition. Faster sequences would be more comfortable for the subject, would increase throughput, and would open up imaging possibilities for three-dimensional imaging and for capturing moving structures in the body. The gradient-echo sequence was developed to be a faster alternative to the spin-echo sequence.

In the gradient-echo sequence, a much smaller flip angle is used than the 90° of the spin-echo sequence. Because a smaller angle is used, it is not necessary to wait a long time (usually around five times the T1 of the tissue of interest when a 90° RF pulse is used) for M_z to recover sufficiently for another repetition. This means that the total acquisition time can be less, leading to the possibility of rapid imaging. The signal measured will be smaller though, because less magnetization is tipped into the $x'y'$ plane to generate a signal. To avoid using a 180° refocusing RF pulse (see "Why Not Use Small-Angle Flips in Spin-Echo Sequences?" box), in the gradient-echo sequence the free induction decay (FID) signal is dephased and then rephased using gradients instead (Figure 10.15).

In the spin-echo approach in which the signal is allowed to dephase and then is refocused with a 180° RF pulse, magnetic field inhomogeneities are eliminated. The gradient-echo approach does not have this effect, and so the inhomogeneities contribute to the lower signal-to-noise ratio for gradient-echo imaging. The faster acquisition is the compensation for the loss of signal-to-noise ratio (SNR).

**WHY NOT USE SMALL-ANGLE FLIPS
IN SPIN-ECHO SEQUENCES?**

Using a small flip angle would not speed up a spin-echo sequence, because the significant M_z component present throughout the sequence would be inverted by the 180° refocusing RF pulse. This would then mean that it was necessary to wait for full recovery of M_z from its inverted position, and consequently the sequence would be even slower than before.

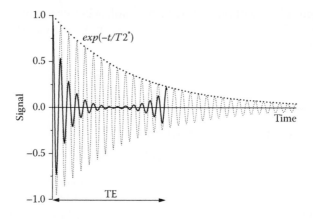

FIGURE 10.15

The dotted lines show the free induction decay (FID) that would occur in the $x'y'$ plane if no dephasing and rephasing gradients were applied. The solid line illustrates the effect on the signal of the gradient-echo sequence gradients. The effect of the dephasing gradient is to reduce the signal. The rephasing gradient reverses the effect of the dephasing gradient, so that a signal grows again and can be acquired at time TE. The amplitude of the measured signal is governed by T2* decay, and because shorter TE times may be used in gradient-echo sequences than in spin-echo sequences, the signal is still large enough to measure.

10.4.2 The Gradient-Echo Imaging Pulse Sequence

The gradient-echo imaging pulse sequence is shown in Figure 10.16. The initial pulse flips the magnetization through $\alpha°$ instead of the 90° used in the spin-echo sequence. There is no 180° RF pulse included. Instead, the frequency encoding gradient has an additional negative lobe that will dephase the spins, before an equal but opposite gradient rephases them again. The time between the initial RF pulse and the measurement of the signal is TE, which is the same as in the spin-echo sequence. Similarly, the repetition time TR is the time between the successive excitation RF pulses.

- In the top row, the $\alpha°$ RF pulse is shown repeated at an interval of TR. Unlike the spin-echo sequence, this is the only RF pulse that is used.

- The next row shows the G_{ss}, or slice select, gradient. It is applied at the time of the $\alpha°$ RF pulse, so that only the spins in that selected slice are saturated and can contribute to the signal.

- The phase encoding gradient is applied once per repetition, with a different strength on each occasion.

- The frequency encoding, or readout, gradient is a key feature of the gradient-echo sequence. There is a negative dephasing lobe followed by a positive gradient that brings the spins back

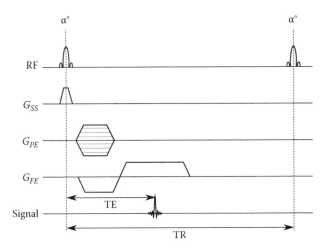

FIGURE 10.16
Simplified timing diagram for the gradient-echo imaging pulse sequence.

into phase to generate the signal and also provides frequency encoding.

- The last row shows when the signal is measured, at a time TE after the α° RF pulse, when rephasing is equal and opposite to the gradient-induced dephasing.

Once again, differently weighted images can be obtained (Table 10.2). As in spin-echo imaging, T1-weighted images are acquired when both TR and TE are short, but for gradient-echo imaging a short TR may have a value under 50 ms. For T1-weighting, the flip angle should be large to help maximize T1 differences (Figure 10.17a). Pure T2-weighting is not possible, because magnetic field inhomogeneities contribute to T2 decay, but T2*-weighting is possible. As in spin-echo imaging, the TE should be long; however, a short TR may be used providing the flip angle is also small, so that full recovery takes place at each repetition (Figure 10.17b). Proton density weighting is achieved using the short TR and small flip angle used for T2*-weighting, but with a short TE to reduce the influence of T2* (Figure 10.17c).

TABLE 10.2

Timing Parameters and Weighting in the Gradient-Echo Imaging Sequence

	TR	TE	Flip Angle
T1-weighted image	Short	Short	Large
T2*-weighted image	Short	Long	Small
Proton density-weighted image	Short	Short	Small

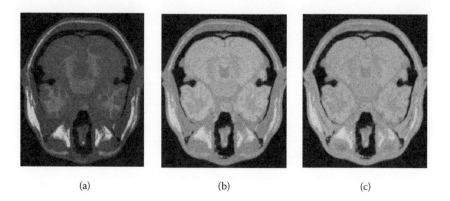

(a)   (b)   (c)

FIGURE 10.17
Examples of images generated using a spoiled (or incoherent) gradient-echo imaging pulse sequence. (a) T1-weighted image with TR = 50 ms, TE = 10 ms, and flip angle = 70°. (b) T2*-weighted image with TR = 50 ms, TE = 20 ms, and flip angle = 5°. (c) Proton density-weighted image with TR = 50 ms, TE = 10 ms, and flip angle = 5°. (Image data from the BrainWeb Simulated Brain Database [1,2]. With permission.)

A comparison between spin-echo imaging and gradient-echo imaging is shown in Table 10.3. Gradient-echo acquisitions are preferred where speed is important, for three-dimensional and dynamic imaging, for example. We shall see later that gradient-echo sequences are used to image flowing blood.

10.4.3 The Ernst Angle

The signal amplitude in gradient-echo imaging depends on TR, TE, and flip angle, and also on the T1 value for the tissue. For any T1 value, at a given TR (<<T1), there is a flip angle that will result in the maximum signal for that tissue in a gradient-echo sequence. This angle is known as the Ernst angle, α_{Ernst}. It can be shown that

$$\cos \alpha_{Ernst} = \exp\left(-\frac{TR}{T1}\right) \tag{10.1}$$

Knowledge of the Ernst angle can be used to maximize the signal from a tissue of interest.

Question 10.7

Use the values for relaxation times given in Table 4.2 for fat, gray matter, white matter, and CSF. Calculate the Ernst angle for each of fat, gray matter, white matter, and CSF when TR = 50 ms.

Answer

The calculated Ernst angles appear in Table 10.4.

TABLE 10.3

A Comparison of the Spin-Echo and Gradient-Echo Imaging Sequences

	Spin-Echo Imaging Sequence	**Gradient-Echo Imaging Sequence**
Flip angle	90° flip angle	Flip angles smaller than 90° may be used
Method used to refocus spins in transverse plane	180° RF pulse	Gradient field
TR and TE	Longer than for gradient echo	Shorter than for spin echo
Overall speed	Slower than gradient echo	Faster than spin echo because shorter TR and TE may be used
Image contrast	Dependent on proton density, tissue relaxation properties, TR, and TE	Dependent on proton density, tissue relaxation properties, TR, TE, and flip angle
SNR	Higher than gradient echo	Decreased signal from small flip angle and short TR, so decreased SNR compared with spin echo
T2-weighting	T2-weighting requires long TR	Can only achieve T2*-weighting, but can be achieved with a short TR and a small flip angle
Image artifacts	Effects of nonrandom inhomogeneity removed by use of 180° RF pulse for refocusing	Prone to artifacts because effects of nonrandom inhomogeneity are not removed by use of gradient for refocusing
		Susceptibility and chemical shift artifacts are a particular problem

TABLE 10.4

Feedback on Question 10.7: The Ernst Angle for Fat, Gray Matter, White Matter, and CSF

	T1/ms	Ernst Angle/° at TR = 50 ms
Fat	260	34.4
Gray matter	920	18.7
White matter	780	20.3
CSF	3,000	1.0

10.4.4 The Steady State

The steady state is a condition that arises in the gradient-echo sequence in which the value of the longitudinal component of magnetization, M_z, has the same value at the start of every TR period. The steady state is

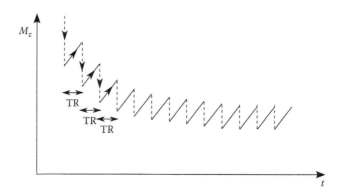

FIGURE 10.18
The steady state is reached after several TR periods. In the steady state, the amount of magnetization flipped out of the z direction by the RF pulse on each repetition (solid lines) is equal to the amount of recovery during each TR period (dashed lines). The magnitude of the steady-state magnetization depends on TR, flip angle, and T1.

reached after several TRs, and it arises because of the following sequence of events, which is illustrated in Figure 10.18. First, in the short TR period after each flip, longitudinal recovery takes place. This means that M_z increases. For the first few repetitions, the amount of magnetization that is recovered each cycle is less than the amount of magnetization that is being flipped out of the z direction by the RF pulse. As a result, the value of M_z goes down with each TR. This reduction in M_z, however, means that less magnetization is flipped at each repetition, so the difference between the amount flipped and the amount recovered (which does not change) becomes less. There comes a point where the amount of M_z flipped at each repetition is exactly the same as the amount that is recovered at each repetition. This means that the system is now in a steady state: M_z no longer gets less at each repetition because it is returned to exactly the same value each time by the amount recovered.

The magnitude of the steady-state value of M_z is highest for short T1 materials, is greater for longer TR values, and gets smaller with increasing flip angle.

For a flip angle α, the components of the net magnetization vector M may be written

$$M_z = M \cos(\alpha)$$

$$M_{xy} = M \sin(\alpha)$$

The relationship between the components means that if M_z is steady, then M_{xy} must also be in a steady state.

We have been concentrating on the longitudinal component of magnetization and the FID signal that arises from that when it is flipped into the transverse plane. However, there is also a transverse component of magnetization arising from the steady state that can give rise to a signal. This component is known as the *residual transverse magnetization*. There are two types of gradient-echo imaging sequence, and they are distinguished by their treatment of the residual transverse magnetization. The first type, described as *coherent* or *rewound* sequences, employs a gradient to ensure that the residual transverse magnetization is rephased, and the residual adds to the signal from the longitudinal component. In the second type, described as *incoherent* or *spoiled* sequences, the residual transverse magnetization is dephased so that the residual does not affect the signal. Note that the images in Figure 10.17 are from a coherent sequence.

The effect of repeated small angle flips on tissue is to reduce the value of M_z to the steady-state value (Figure 10.18). If tissue enters an imaging field that has not been subject to the repeated RF pulses, it will give a higher signal. This is known as the *inflow effect*, and we will consider it in more detail in Section 10.5.2.

10.5 Flow Phenomena

In our consideration of the imaging sequences so far, it has been assumed that the subject of imaging is not moving. Blood flowing during the acquisition of an image does not satisfy this assumption, and can have a major effect on the appearance of the image. There are two categories of effect: *time-of-flight effects* and *phase effects*. Time-of-flight effects arise because blood that was present in the selected slice at the start of the imaging sequence may no longer be present in the slice for the later stages of the imaging sequence. Phase effects arise because different velocities lead to blood being affected by different field strengths, with consequent dephasing.

10.5.1 Time-of-Flight Flow Phenomenon in Spin-Echo Imaging

The time-of-flight flow effect is illustrated for the spin-echo imaging sequence in Figure 10.19. Time-of-flight effects also occur with the gradient-echo imaging sequence, but the differences between the imaging sequences mean that the results are different. There are three rows of diagrams in Figure 10.19. In the top row, flow is fast; in the middle row, flow is slow; and in the bottom row, there is no flow at all. Each row shows three time points for the spin-echo sequence. At the left is the time when the initial 90° RF pulse is applied, the center image is at the time when the 180° RF refocusing pulse is applied (TE/2), and the right-hand image is at the time the signal is acquired (TE).

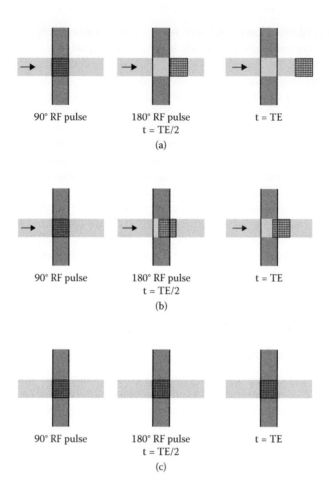

FIGURE 10.19

Flow and the spin-echo sequence. The selected slice is the dark gray area between the strong black lines. A blood vessel is at right angles to the selected slice, shown in lighter gray. The velocity of flow in the blood vessel is different for (a), (b), and (c). (a) Flow is fast in this case, as indicated by the bold arrow. All the blood that received the 90° RF pulse has left the slice at the time of the refocusing 180° RF pulse. As no blood has received both pulses, no signal will be obtained from the blood vessel under these circumstances, and it will appear dark in an image. This is washout. (b) Flow is slower than in (a), indicated by the thin arrow. As a result, not all the blood that received the 90° RF pulse has left the slice when the refocusing 180° RF pulse is applied. The signal received depends on the volume of blood that has received both RF pulses, and so the blood vessel will no longer be signal-free in the image. (c) For comparison, there is no flow. All the blood receives both RF pulses, and the image will show the usual signal expected for stationary blood. In all three cases, the location of the blood at the time the measurement is made (*t* = TE) does not affect the signal. The demonstrator program (Section 10.6) reproduces the diagrams for the three different flow velocities, and allows images to be generated for different slice thicknesses and values of TE.

In the first row (fast flow), none of the blood that was present in the slice at the time of the 90° RF pulse is still in the slice when the 180° refocusing RF pulse is applied. In the spin-echo sequence a signal is only possible if tissue receives both 90° and 180° RF pulses, so the image associated with this situation will be dark in areas where flow is present.

In the bottom row, there is no flow at all. This means that all the blood in the imaged slice that received the 90° RF pulse is still present to receive the 180° refocusing RF pulse and will emit a signal. The image associated with this situation will be T1-weighted or T2-weighted in the usual way, depending on the sequence parameter values.

In the middle row, where the flow is slower than in the top row, not all of the flowing blood that received the 90° RF pulse has left the slice at the time of the 180° refocusing RF pulse. The blood that is still present will receive both pulses and emit a signal. This signal will come from a smaller volume of blood than is the case when there is no flow, leading to a reduction in signal intensity in the vessel compared with the no-flow case.

Therefore, in the spin-echo imaging sequence we have seen that flow can lead to a reduction in signal or complete loss of signal compared to the no-flow case.

10.5.1.1 Condition for Signal Loss in Spin-Echo Imaging

Signal loss will occur if the flow velocity is sufficient for all the blood present at the time of the 90° RF pulse to have left the slice after a time TE/2. Using the notation of Chapter 9, where slice thickness is *Pixel size*$_{SS}$, the condition can be expressed as

$$v_{signal\ loss} \geq \frac{2\ Pixel\ size_{SS}}{TE} \tag{10.2}$$

Images acquired using values for the slice thickness and TE that result in signal loss from most of the blood flowing through the slice are called *black blood* images. The phenomenon is sometimes called *washout*.

10.5.2 Time-of-Flight Flow Phenomenon in Gradient-Echo Imaging

The time-of-flight flow effect is illustrated for the gradient-echo imaging sequence in Figure 10.20. The effects are different for the gradient-echo imaging sequence because there is no refocusing RF pulse in the sequence. This means that the requirement for the blood to be present in the slice at the time of both the initial RF pulse and the later pulse does not apply. Instead, the significant point is whether or not the blood has completely left the slice during the TR period. If all the blood has left the

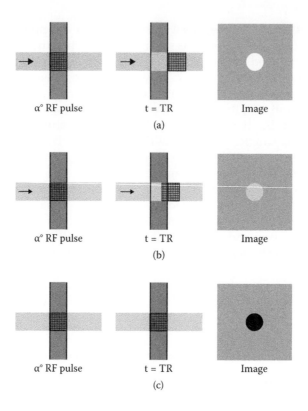

FIGURE 10.20

Flow and the gradient-echo sequence. The selected slice is the dark gray area between the strong black lines. A blood vessel is at right angles to the selected slice, shown in lighter gray. The velocity of flow in the blood vessel is different for (a), (b), and (c). (a) Flow is fast in this case, as indicated by the bold arrow. At time t = TR, when the next $\alpha°$ RF pulse is applied, all the blood that had received the previous $\alpha°$ RF pulse has left the slice, and has been replaced by blood where the spins have a larger M_z component, as they have not previously been subject to any RF pulses. This means that a large signal is obtained from the blood vessel, a phenomenon that is called the inflow effect. The image on the right shows how the strong signal would appear as bright blood in an image. (b) Flow is slower than in (a), indicated by the thin arrow. As a result, not all the blood that received the previous $\alpha°$ RF pulse has left the slice, as only a proportion has been replaced by spins not previously subject to any RF pulses. This means that the signal from the blood vessel is strong, but not as strong as when all of the blood is replaced. This is illustrated in the image at the right. (c) For comparison, there is no flow. All the blood remains in the slice and, like all the stationary tissue, is subject to repeated RF pulses. The repeated RF pulses lead to a smaller M_z component than for tissue that receives just one RF pulse. The signal from the blood vessel is lower than when flow is present, as illustrated by the image at the right.

slice, at the time of the next RF flip the flowing spins present in the slice will not have experienced any of the previous RF pulses. Repeated RF pulses in the gradient-echo sequence have the effect of reducing the signal (Section 10.4.4), so the incoming flowing spins will give a larger signal than spins that remain in the slice. This is illustrated in Figure 10.20. In the top row, flow is fast; in the middle row, flow is slow; and in the bottom row, there is no flow at all. Each row shows two time points for the gradient-echo sequence. At the left is the time when the initial $\alpha°$ RF pulse is applied, and the right-hand image is at time TR later, when the next $\alpha°$ RF pulse is applied.

In the first row (fast flow), none of the blood that was present in the slice at the time of the first $\alpha°$ RF pulse is still in the slice at a time TR later when the next $\alpha°$ RF pulse is applied. All of the flowing spins in the slice have been replaced by inflowing spins that have not experienced previous RF pulses. The image associated with this situation will be bright in areas where flow is present.

In the bottom row, there is no flow at all. This means that all the blood in the imaged slice that was present in the slice at the time of the first $\alpha°$ RF pulse is still in the slice at TR when the next $\alpha°$ RF pulse is applied. The signal is at the usual steady-state level seen in the gradient-echo sequence, which is lower than the signal from inflowing blood. In the middle row, where the flow is slower than in the top row, not all of the flowing blood that was present in the slice at the time of the first $\alpha°$ RF pulse has left it again when the next RF pulse is applied. Again, the inflowing blood will lead to a large signal, but this signal will come from a smaller volume of blood than is the case when there is complete replacement of spins, leading to a reduction in signal intensity in the vessel compared with the fast-flow case.

Therefore, in the gradient-echo imaging sequence we have seen that flow can lead to an increase in signal compared to the no-flow case.

10.5.2.1 Condition for the Maximum Signal in Gradient-Echo Imaging

The maximum signal will occur if the flow velocity is sufficient for all the blood present at the time of the $\alpha°$ RF pulse to have left the slice after a time TR. Using the notation of Chapter 9, where slice thickness is *Pixel size*$_{SS}$, the condition can be expressed as

$$v_{max\,signal} \geq \frac{Pixel\,size_{SS}}{TR} \tag{10.3}$$

Images acquired using values for the slice thickness and TR that result in a high signal from the blood flowing through the slice are called *bright blood* images.

10.5.3 Phase Effects with Flow

All imaging sequences include the use of magnetic field gradients for spatial localization. The phase encoding gradient is designed to introduce the phase shifts required for localization, and care is taken to remove phase differences introduced by the slice selection and frequency encoding gradients. However, the gradient designs work correctly only for static tissue. Any moving spin will acquire an unplanned phase shift, and the size of that phase shift will depend on the velocity.

10.5.3.1 Intravoxel Effects Leading to Signal Loss

Flow velocity is not uniform throughout the vessel; for example, there is stationary blood close to the vessel walls with the fastest flow in the center, so each voxel within flowing blood will usually contain a range of different velocities. Within each voxel the range of different phase shifts can be large enough for complete dephasing to take place, leading to signal loss.

10.5.3.2 Pulsatile Flow and Ghosting

In pulsatile flow, the blood velocity changes with time, and the velocity may differ from TR period to TR period. In this case, even when the flow velocity is uniform throughout the vessel, the phase shifts acquired by flowing spins will differ for each TR. The usual reason that differing phase shifts with each TR period arise is as a result of phase encoding, and so phase shifts arising from pulsatile flow are interpreted as encoding information about spatial location. The result is the appearance of ghost images of the vessel, offset in the phase encoding direction from the true location.

10.6 Flow Phenomena Demonstrator: Spin-Echo Imaging Sequence

Start the Flow Phenomena program as explained in Chapter 1. The main part of the window (Figure 10.21) is laid out in the same way as Figure 10.19. Flow is fast in the top row, medium sized in the middle, and there is no flow in the bottom row. There are three columns, also similar to Figure 10.19, each associated with a time point in the spin-echo sequence. The far-right column shows simulated spin-echo images acquired at right angles to the direction of flow. In the demonstrator, TR is fixed at 2,000 ms and TE can be changed using the slider at the bottom left of the window. When the program starts, TE is set to 50 ms and the slice thickness is 5 mm. Leave these settings as they are. The three flow velocities are fixed in the demonstrator, but we shall see that the appearance of the associated image changes depending on TE and the slice thickness.

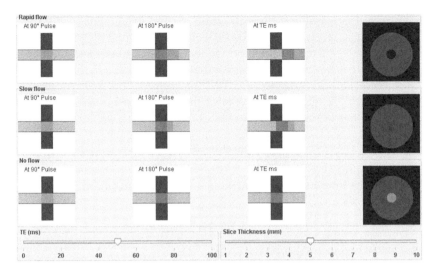

FIGURE 10.21

The appearance of the Flow Phenomena demonstrator (spin-echo imaging sequence) when the program is started. The diagrams are in color in the demonstrator. Blue shading represents the blood that received only the 90° RF pulse; green shading is blood that received the 180° refocusing RF pulse but did not first receive the 90° RF pulse; red shading represents the blood that received both pulses.

The demonstrator is designed to start up with the acquisition parameters set so that the fast-flow velocity in the top row is just fast enough to ensure washout from the 5-mm slice in the time period TE/2. The same flow velocity will give an image with a different appearance if the acquisition parameters are changed. Use the slider at the bottom right to change the slice thickness from 5 mm to 10 mm. Blood would now have to flow twice as far to be completely washed out of the slice before the 180° refocusing RF pulse. As a result, with the thicker slice, a signal is received so the lumen of the blood vessel is no longer black in the image. There has not been complete washout.

Return the slice thickness to 5 mm. Consider the middle row, where the flow is slower than in the top row. Do you expect the lumen in the image to get brighter or darker when the slice thickness is increased? What do you expect to see if the slice thickness is decreased? Check your predictions using the demonstrator. You should note that even with the slower flow, complete washout, resulting in a lack of signal in the image, occurs for thin slices, as would also be expected from Equation 10.2.

Return the slice thickness to 5 mm and TE to 50 ms. In the fast-flow row this is the situation corresponding with the condition in Equation 10.2, and signal loss occurs. There is no red shading in the vessel in the center diagram of the fast-flow row at the time of the 180° refocusing RF pulse. Think about what change is to be expected if the value of TE is reduced. There is now less time for the blood to leave the slice before the 180° refocusing RF pulse. So, we would expect that as TE is reduced, the brightness of the vessel lumen will increase, because more blood that has received the 90° RF pulse remains in the slice. In the demonstrator, we see red shading appear in the vessel, and the lumen of the image gets brighter.

Question 10.8

Change the slice thickness to 7 mm, then adjust TE using the slider until the top row is at the signal loss boundary conditions for this thicker slice.

Answer

Signal loss should happen when TE ≥ 70 ms.

Question 10.9

Set TE to 60 ms. Adjust the slice thickness slider to determine the slice thickness below which signal loss will occur for the middle row in the demonstrator, which has a slower flow than the top row.

Answer

Signal loss should happen when the slice thickness is less than 3 mm.

Question 10.10

The velocity of venous flow in a vessel is 15 cm s⁻¹. The vessel is perpendicular to a slice thickness of 4 mm. Which of the following values of TE would lead to complete signal loss in the vessel when using a spin-echo sequence?

   (a) 25 ms
   (b) 55 ms
   (c) 105 ms
   (d) 45 ms
   (e) 75 ms

Answer

Equation 10.2 may be used to calculate that signal loss occurs for values of TE ≥ 53 ms. So (b), (c), and (e) are all correct.

Question 10.11

The velocity of venous flow in a vessel is 15 cm s⁻¹. The vessel is perpendicular to a slice thickness of 4 mm. What is the smallest TR that would lead to the maximum signal from the vessel when using a gradient-echo sequence?

Answer

Substituting into Equation 10.3 leads to an answer of approximately 27 ms.

Question 10.12

Which of the following statements are true concerning the gradient-echo sequence and flow across the slice?

(a) Signal loss occurs for rapidly flowing blood.
(b) Signal enhancement occurs for rapidly flowing blood.
(c) Signal loss occurs for slowly flowing blood.
(d) Signal enhancement occurs for slowly flowing blood.
(e) Signal enhancement is more likely to occur for longer TR.

Answer

The true statements are (b), (d), and (e).

Question 10.13

Phase effects arising from flow within a vessel can have which of the following results?

(a) Signal loss in the image of the vessel.
(b) Signal enhancement in the image of the vessel.
(c) Signal appearing in the wrong place in an image.
(d) Ghosting.
(e) Vessels in bright blood images appearing smaller than their true size.

Answer

All the statements are true. Signal loss from intravoxel dephasing near the vessel walls could lead to underestimation of the vessel diameter as suggested in (e).

10.7 Magnetic Resonance Angiography (MRA)

Magnetic resonance angiography (MRA) techniques are designed to show blood vessels with a high signal while minimizing the signal acquired from other tissue. In all cases the data are visualized using the image processing technique of maximum intensity projection (MIP). MIP works well on image data where the object of interest is bright and the signal from other tissue is low, which is the situation in MRA. The result of MIP postprocessing is a series of images that have the appearance of projection images acquired at different viewpoints around the volume (the stack of acquired slices). When run as a movie sequence, the images give the impression of three dimensions and highlight the spatial relationships between the vessels.

The MRA sequences make use of the time-of-flight and phase effects previously described, and fall into three categories: time-of-flight angiography, phase-contrast angiography, and contrast-enhanced angiography.

10.7.1 Time-of-Flight Angiography

Angiographic images are acquired using gradient-echo sequences that have a bright signal from flowing blood because of the inflow effect. To reduce the signal from stationary tissue, a short TR is chosen so that full recovery of M_z does not take place, together with a midrange flip angle and short TE. Two different types of acquisitions can be used: two-dimensional or three-dimensional. A two-dimensional acquisition is the kind of acquisition that we have considered so far, where a set of thin slices is obtained, which may then be stacked together to represent a three-dimensional volume. A three-dimensional acquisition involves an extra phase encoding step. A thick slab of tissue is excited (instead of a slice at a time), and the individual slices within that slab are defined using the second phase encoding step in the slice selection direction. Three-dimensional acquisitions lead to thinner slices than two-dimensional acquisition, and the images have a higher SNR. However, they do not show slow flow as well as two-dimensional acquisitions, and the inflow effect is less strong because of the larger volume of the slab. Time-of-flight (TOF) angiography works best for fast flow and when flow is at right angles to the selected slice, because flow through the plane of the slice does not demonstrate the inflow effect.

10.7.2 Phase-Contrast Angiography

The phase phenomena described in the previous discussion of flow phenomena all resulted in signal loss. The images acquired using phase-contrast MRA, however, have bright blood because of the way in which the effects of phase are harnessed. The phase-contrast MRA technique involves applying specially designed gradients. The gradients ensure that the velocity of the blood is directly related to the phase of the MR signal, in a process called *velocity encoding*. If the maximum velocity expected is known, then the gradients can be set up so that no phase shift is greater than 180°. The velocity encoding gradients are applied twice, and for the second application their direction is reversed compared with the first. The two sets of phase images are calculated, and then one set is subtracted from the other. On subtraction, the constant background from stationary material will be set to zero, while positive and negative values will be left where flow is present. The technique is directionally sensitive, which means that the velocity encoding gradients must be applied in all three directions to capture information about blood flow in any direction. Phase-contrast imaging can be used to produce both numerical estimates of blood flow velocity and angiograms. As for the time-of-flight methods, both two- and three-dimensional acquisitions are possible.

10.7.3 Contrast-Enhanced Angiography

The time-of-flight method is compromised when the inflow of fresh spins required for the bright signal cannot be achieved, perhaps because of the direction of flow or a large field of view (FOV). In this case, contrast agents that reduce T1 to below that of their surroundings, and so appear bright in T1-weighted images, can be used. The contrast agent is injected as a bolus, and images are acquired using a T1-weighted gradient-echo sequence. In contrast-enhanced angiography the appearance of the vessels depends on the presence of the contrast agent and not on flow phenomena, so problems associated with flow within the slice plane and with saturation of spins do not arise.

Question 10.14

Insert the correct MRA technique in each of the following sentences regarding magnetic resonance angiography (MRA).

(a) _____ MRA is directionally sensitive.
(b) _____ MRA involves the use of additional gradients.
(c) _____ techniques provide quantitative information about flow speed.
(d) _____ techniques rely on the shortening of the T1 of blood by a contrast agent.
(e) _____ MRA works best with high-velocity flow.

Answer

The missing words are (a) phase-contrast, (b) phase-contrast, (c) phase-contrast, (d) contrast-enhanced, and (e) time-of-flight.

10.8 Chapter Summary

In this chapter two further imaging sequences were introduced. The inversion recovery imaging sequence is closely related to spin-echo imaging, but the use of an initial inversion RF pulse means that it is possible to choose to null the signal from material with a particular T1 value. Interactive tools were used to reinforce key points associated with inversion recovery imaging and tissue nulling. In gradient-echo imaging a gradient is used deliberately to dephase and then rephase the signal, and this means that much faster acquisitions are possible than with spin-echo imaging. There are, of course, many more imaging sequences used in MRI. For any new imaging sequence, however, the underlying principles that we have used to understand the working of spin-echo imaging, inversion recovery imaging, and gradient-echo imaging will be the same. The effects of flow on spin-echo and gradient-echo images were covered next, taking advantage of an interactive diagram to emphasize the relationships between the parameters associated with signal loss in the spin-echo sequence. The chapter closed with an overview of imaging sequences designed to demonstrate flowing blood.

References

1. The BrainWeb Simulated Brain Database. www.bic.mcgill.ca/brainweb/.
2. Collins, D. L., et al. 1998. Design and construction of a realistic digital brain phantom. *IEEE Trans. Med. Imag.* 17:463.

11

Multiple-Choice Questions

This chapter contains 150 multiple-choice questions and answers. The questions cover all the material presented in the book, but there are relatively more questions on the background topics of Chapter 2, to provide a revision tool if required.

The questions are arranged so that there are groups of three related questions. This arrangement means that the reader can either focus on a topic of interest or divide the long list of questions into three tests of 50 questions each and still cover all the material of the book. To tackle subtests, use every third question, that is, questions 1, 4, 7, 10, ... for the first test; questions 2, 5, 8, 11, ... for the second test, and questions 3, 6, 9, 12, ... for the third test.

Many of the questions have more than one correct answer. Where there is only one correct answer, this is indicated with the symbol § at the beginning of the question.

11.1 Multiple-Choice Questions

1. § The tangent of the angle X shown in Figure 11.1 is given by:
   - ☐ (a) b/c
   - ☐ (b) b/a
   - ☐ (c) a/c
   - ☐ (d) a/b
   - ☐ (e) c/a

2. § The sine of the angle X shown in Figure 11.1 is given by:
   - ☐ (a) b/c
   - ☐ (b) b/a
   - ☐ (c) a/c
   - ☐ (d) a/b
   - ☐ (e) c/a

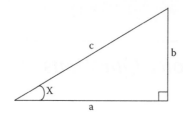

FIGURE 11.1
Diagram for multiple-choice questions 1 to 3.

3. § The cosine of the angle X shown in Figure 11.1 is given by:
☐ (a) b/c
☐ (b) b/a
☐ (c) a/c
☐ (d) a/b
☐ (e) c/a

4. § The tangent of the angle X shown in Figure 11.2 is given by:
☐ (a) b/c
☐ (b) b/a
☐ (c) a/c
☐ (d) a/b
☐ (e) c/a

5. § The sine of the angle X shown in Figure 11.2 is given by:
☐ (a) b/c
☐ (b) b/a
☐ (c) a/c
☐ (d) a/b
☐ (e) c/a

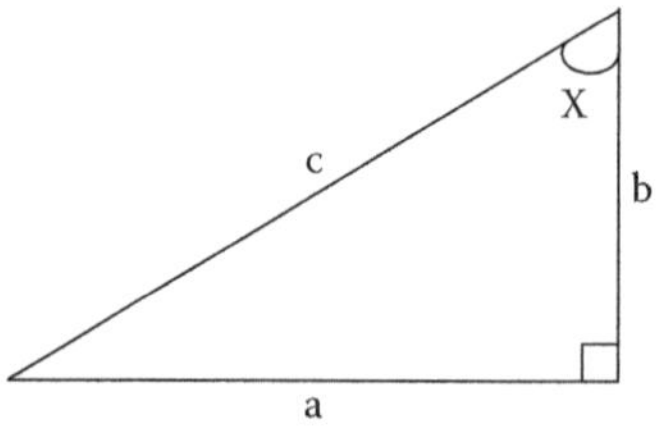

FIGURE 11.2
Diagram for multiple-choice questions 4 to 6.

6. § The cosine of the angle X shown in Figure 11.2 is given by:
   ☐ (a) b/c
   ☐ (b) b/a
   ☐ (c) a/c
   ☐ (d) a/b
   ☐ (e) c/a

7. $\pi/2$ radians is equivalent to:
   ☐ (a) $180°$
   ☐ (b) $270°$
   ☐ (c) The phase difference between $\sin(x)$ and $\cos(x)$
   ☐ (d) $90°$
   ☐ (e) $360°$

8. Which of the following statements are true?
   ☐ (a) $360° = 2\pi$ radians
   ☐ (b) $90° = \pi/2$ radians
   ☐ (c) 1 radian $= 180°$
   ☐ (d) 10π radians $= 1{,}800°$
   ☐ (e) $360° = \pi$ radians

9. Which of the following statements are true?
   ☐ (a) $180° = 2$ radians
   ☐ (b) $180° = 2\pi$ radians
   ☐ (c) $180° = \pi$ radians
   ☐ (d) 4π radians $= 360°$
   ☐ (e) $\pi/2$ radians $= 90°$

10. Which of the following statements are true?
    ☐ (a) An angle of 1 radian is about $57°$ in size.
    ☐ (b) An angle of 1 radian is about $3.142°$ in size.
    ☐ (c) When $x = 1$ radian, $\sin(x) = 0$.
    ☐ (d) When $x = 1$ radian, $\sin(x) = 1$.
    ☐ (e) When $x = \pi$ radians, $\sin(x) = 0$.

11. Which of the following statements are true?
    ☐ (a) When $x = 0$ radians, $\sin(x) = 0$.
    ☐ (b) When $x = \pi$ radians, $\sin(x) = 0$.
    ☐ (c) When $x = 2\pi$ radians, $\sin(x) = 0$.
    ☐ (d) When $x = 3\pi$ radians, $\sin(x) = 0$.
    ☐ (e) When $x = 4\pi$ radians, $\sin(x) = 0$.

12. Which of the following statements are true?
- ☐ (a) When $x = 0$ radians, $\cos(x) = 0$.
- ☐ (b) When $x = \pi/2$ radians, $\sin(x) = 0$.
- ☐ (c) When $x = 2\pi$ radians, $\cos(x) = 1$.
- ☐ (d) When $x = \pi$ radians, $\sin(x) = 1$.
- ☐ (e) When $x = 3\pi/2$ radians, $\cos(x) = 0$.

13. Vectors:
- ☐ (a) Have amplitude
- ☐ (b) Have direction
- ☐ (c) Can be resolved into components
- ☐ (d) Cannot be added
- ☐ (e) Have phase

14. Vectors:
- ☐ (a) Have both amplitude and direction
- ☐ (b) Have a direction only
- ☐ (c) Can also be called scalars
- ☐ (d) Can be added by summing the vector amplitudes
- ☐ (e) Can be added by a process that involves first separately summing the x and y components

15. The following are vector quantities:
- ☐ (a) Velocity
- ☐ (b) Displacement
- ☐ (c) Speed
- ☐ (d) Distance
- ☐ (e) Magnetic field strength

16. For the vector shown in Figure 11.3, and assuming that phase is measured in an anticlockwise direction from the positive x axis, the following are true:
- ☐ (a) x component = −6.69
- ☐ (b) x component = 6.02
- ☐ (c) y component = 6.02
- ☐ (d) Phase = 48°
- ☐ (e) Phase = 138°

17. For the vector shown in Figure 11.3, and assuming that phase is measured in an anticlockwise direction from the positive x axis, the following are true:
- ☐ (a) Phase = 42°
- ☐ (b) Phase = 228°

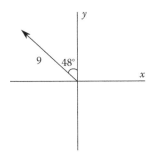

FIGURE 11.3
Diagram for multiple-choice questions 16 and 17.

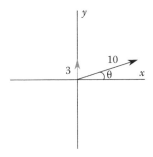

FIGURE 11.4
Diagram for multiple-choice question 18.

☐ (c) Phase = 138°

☐ (d) y component = −6.02

☐ (e) y component = 6.02

18. For the vector shown in Figure 11.4, the following are true:

☐ (a) The y component of the vector shown is 3.

☐ (b) The y component of the vector shown is 10sinθ.

☐ (c) The x component of the vector shown is 10.

☐ (d) The x component of the vector shown is 10cosθ.

☐ (e) The amplitude of the vector shown is 3.

19. With reference to the graph in Figure 11.5, where τ is the time constant, which of the following statements are true?

☐ (a) The time constant for the solid black curve is less than the time constant for the dotted black curve.

☐ (b) The time constant for the dotted black curve is less than the time constant for the solid gray curve.

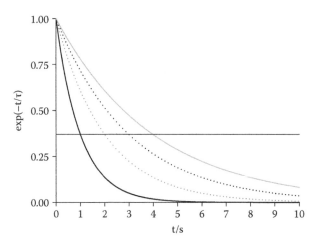

FIGURE 11.5

Diagram for multiple-choice questions 19 to 21. τ is the time constant, which is different for each of the exponential decay curves shown. The horizontal line is at the value 0.37.

☐ (c) The time constant for the solid black curve is greater than the time constant for the dotted black curve.

☐ (d) The time constant for the dotted black curve is 5 seconds.

☐ (e) The time constant for the dotted black curve is 3 seconds.

20. With reference to the graph in Figure 11.5, where τ is the time constant, which of the following statements are true?

☐ (a) The curve with the shortest time constant is the solid black one.

☐ (b) The curve with the shortest time constant is the solid gray one.

☐ (c) The time constant for the dotted gray curve is 2 seconds.

☐ (d) The time constant for the dotted gray curve is 6 seconds.

☐ (e) The time constant for the dotted gray curve is 0.37 second.

21. With reference to the graph in Figure 11.5, where τ is the time constant, which of the following statements are true?

☐ (a) The time constant for the solid black curve is 5 seconds.

☐ (b) The time constant for the solid black curve is 1 second.

☐ (c) The time constant for the solid black curve is 0.37 second.

☐ (d) The time constant for the solid black curve is less than the time constant for the dotted gray curve.

☐ (e) The time constant for the dotted gray curve is greater than the time constant for the dotted black curve.

22. Which of the following statements regarding notation are true?
    - ☐ (a) The square root of a number A may be written as $A^{1/2}$.
    - ☐ (b) The square root of a number A may be written as $A^{-1/2}$.
    - ☐ (c) The square root of a number A may be written as A^2.
    - ☐ (d) The square root of a number A may be written as $\sqrt{A}$.
    - ☐ (e) The square root of a number A may be written as $\exp(A)$.

23. Which of the following statements regarding notation are true?
    - ☐ (a) The reciprocal of a number B may be written as $1/B$.
    - ☐ (b) The reciprocal of a number B may be written as B^{-1}.
    - ☐ (c) The reciprocal of a number B may be written as $\frac{1}{B}$.
    - ☐ (d) The reciprocal of a number B may be written as $B^{1/2}$.
    - ☐ (e) The reciprocal of a number B may be written as $\sqrt{B}$.

24. Which of the following statements regarding notation are true?
    - ☐ (a) $y = \exp(-A)$ represents exponential decay.
    - ☐ (b) $y = A^{-e}$ represents exponential decay.
    - ☐ (c) $y = A^{1/2}$ is the square root of A.
    - ☐ (d) $y = \exp(A)$ represents exponential decay.
    - ☐ (e) $y = A^{-1}$ represents exponential growth.

25. FFT:
    - ☐ (a) Stands for Frequency Fourier Transform
    - ☐ (b) Stands for Fast Frequency Transfer
    - ☐ (c) Stands for Fast Fourier Transform
    - ☐ (d) Provides a frequency spectrum of a signal
    - ☐ (e) Gives data that can sometimes be easier to work with than the untransformed data

26. The FFT:
    - ☐ (a) Requires the number of data points to be a multiple of 2
    - ☐ (b) Requires the number of data points to be 255
    - ☐ (c) Is a computational method of performing Fourier analysis
    - ☐ (d) Requires the number of data points to be a power of 2
    - ☐ (e) Is an analytical technique never used on measured data

27. Which of the following statements are true?
    - ☐ (a) Fourier analysis can be performed only on periodic functions.
    - ☐ (b) Fourier analysis allows any function to be expressed as a series of sine and cosine terms, each with a single frequency.

- ☐ (c) Fourier analysis converts a function into a series of cosine terms, each with a single frequency.
- ☐ (d) The sinc function is the Fourier transform of a sine function.
- ☐ (e) The sinc function is the Fourier transform of a square or rectangular pulse.

28. Which of the following statements are true?
    - ☐ (a) The charge on a single electron or proton is $\pm 1.6 \times 10^{-19}$ C.
    - ☐ (b) An atom has the same number of electrons as protons.
    - ☐ (c) The atomic number of an atom, which has the symbol Z, is the number of protons in the nucleus.
    - ☐ (d) The atomic number of an atom, which has the symbol A, is the number of neutrons in the nucleus.
    - ☐ (e) The charge on a single electron is 1.6×10^{-19} C.

29. § The nucleus of an atom is made up of:
    - ☐ (a) Electrons
    - ☐ (b) Neutrons and protons
    - ☐ (c) Protons and electrons
    - ☐ (d) Neutrons and electrons
    - ☐ (e) Protons

30. § Which of the following statements is true?
    - ☐ (a) An atom has no charge.
    - ☐ (b) An ion has no charge.
    - ☐ (c) The charge on an atom is -1.6×10^{-19} C.
    - ☐ (d) The charge on a proton is -1.6×10^{-19} C.
    - ☐ (e) The charge on an ion is called the atomic number.

31. Which of the following statements are true?
    - ☐ (a) The unit of electric current is the coulomb (C).
    - ☐ (b) The unit of electric current is the ampere (A).
    - ☐ (c) The unit of electric potential is the ampere (A).
    - ☐ (d) The unit of electric potential is the volt (V).
    - ☐ (e) The ampere is the same as a coulomb per second.

32. § The letter B is used to represent:
    - ☐ (a) A magnetic field
    - ☐ (b) An electric field
    - ☐ (c) Frequency

- ☐ (d) Current
- ☐ (e) Displacement

33. Which of the following statements are true?
    - ☐ (a) The unit of magnetic field strength is the coulomb (C).
    - ☐ (b) The unit of magnetic field strength is the tesla (T).
    - ☐ (c) The gauss (G) is a non-SI alternative unit for magnetic field strength.
    - ☐ (d) The unit of electric charge is the coulomb (C).
    - ☐ (e) The unit of magnetic susceptibility is T m^{-1}.

34. Which of the following statements are true for paramagnetic materials?
    - ☐ (a) The magnetic susceptibility is less than 0.
    - ☐ (b) The magnetic susceptibility is greater than 1.
    - ☐ (c) Examples include aluminum, manganese, oxygen, platinum, and tungsten.
    - ☐ (d) Examples include iron, cobalt, and gadolinium.
    - ☐ (e) The field within the material is higher than the main field.

35. Which of the following statements are true regarding magnetic susceptibility?
    - ☐ (a) Magnetic susceptibility has the same units as magnetic field strength.
    - ☐ (b) Magnetic susceptibility is the extent to which a material becomes magnetized when placed in a magnetic field.
    - ☐ (c) In a magnetic field of 1 T, magnetic susceptibility is the same for all materials.
    - ☐ (d) Magnetic susceptibility is a dimensionless quantity.
    - ☐ (e) Magnetic susceptibility is the size of the magnetic field outside the main bore of the magnet.

36. Which of the following statements are true for ferromagnetic materials?
    - ☐ (a) The magnetic susceptibility is greater than 1.
    - ☐ (b) The magnetic susceptibility is less than 0.
    - ☐ (c) Examples include bismuth and copper.
    - ☐ (d) Examples include iron, cobalt, and gadolinium.
    - ☐ (e) The induced magnetic field within the material opposes the main field.

37. Radio frequency electromagnetic radiation:
    - ☐ (a) Is nonionizing
    - ☐ (b) Has a longer wavelength than visible light
    - ☐ (c) Has a longer wavelength than x-radiation
    - ☐ (d) Has a higher frequency than infrared radiation
    - ☐ (e) Is used in magnetic resonance imaging (MRI)

38. Regarding electromagnetic radiation, which of the following statements are true?
    - ☐ (a) Electromagnetic radiation consists only of an oscillating electric field (E).
    - ☐ (b) Electromagnetic radiation consists of both an oscillating electric field (E) and an oscillating magnetic field (B).
    - ☐ (c) The velocity of propagation is the velocity of light.
    - ☐ (d) The direction of propagation is the same as the direction of the oscillations of the electric field.
    - ☐ (e) X-rays are not electromagnetic radiation.

39. Regarding electromagnetic radiation, which of the following statements are true?
    - ☐ (a) Electromagnetic radiation consists only of an oscillating magnetic field.
    - ☐ (b) Electromagnetic radiation consists of both an oscillating magnetic field and an oscillating electric field.
    - ☐ (c) Electromagnetic radiation consists of both a static magnetic field and a static electric field.
    - ☐ (d) The direction of propagation of the radiation is in the same direction as the magnetic field.
    - ☐ (e) X-rays and radio frequency radiation are both electromagnetic radiation.

40. In each of the options below there is a word or phrase, followed by a definition. Which are the true statements? If the definition is the correct definition of that word, then the statement is true. All the words are real words, and all the definitions real definitions. The thing to identify is that they match.
    - ☐ (a) *In quadrature*: 90° out of phase.
    - ☐ (b) *Time constant*: Time for an exponential signal to decay to 0.37 of its initial value.
    - ☐ (c) *Electron*: An atom following removal of an electron from an inner orbit.

☐ (d) *Ion*: Small particle carrying a negative charge.

☐ (e) *Ferromagnetic*: Material with magnetic susceptibility greater than one.

41. In each of the options below there is a word or phrase, followed by a definition. Which are the true statements? If the definition is the correct definition of that word, then the statement is true. All the words are real words, and all the definitions are real definitions. The thing to identify is that they match.

☐ (a) *Proton*: Small particle found in the nucleus carrying a positive charge.

☐ (b) *Coulomb*: Unit of electrical charge.

☐ (c) *Paramagnetic*: A pair of magnetic poles, one north pole and one south pole.

☐ (d) *Eddy current*: Current induced by a changing magnetic field in the iron core of a magnet.

☐ (e) *Magnetic dipole*: Material with magnetic susceptibility greater than zero.

42. In each of the options below there is a word or phrase, followed by a definition. Which are the true statements? If the definition is the correct definition of that word, then the statement is true. All the words are real words, and all the definitions are real definitions. The thing to identify is that they match.

☐ (a) *Tesla*: Small particle found in the nucleus carrying no charge.

☐ (b) *Atomic number*: Number of protons in the nucleus of an atom.

☐ (c) *RF*: Abbreviation for radio frequency.

☐ (d) *Neutron*: The number of cycles per second measured in hertz.

☐ (e) *Linear frequency*: Unit of magnetic field strength.

43. § 1.0 tesla is equal to how many gauss?

☐ (a) 1,000

☐ (b) 2,000

☐ (c) 10,000

☐ (d) 20,000

☐ (e) 5,000

44. Which of the following statements are true?

☐ (a) 0.001 T = 1 mT

☐ (b) 1,000 T = 1 mT

☐ (c) 1,000 µT = 1 mT
☐ (d) 10^{-3} T = 1 mT
☐ (e) 10^3 T = 1 mT

45. Which of the following statements are true?

☐ (a) 1 G = 10^{-4} T
☐ (b) 0.1 mT = 1 G
☐ (c) 0.1 mT = 0.0001 T
☐ (d) 100 G = 0.01 T
☐ (e) 0.01 T = 10^{-2} T

46. Which of the following statements are true?

☐ (a) Image display monitors usually display 8 bits per pixel.
☐ (b) Image display monitors usually display 16 bits per pixel.
☐ (c) There are 256 different gray levels available on a 8-bit display.
☐ (d) Eight-bit displays cannot display images that have a matrix size larger than 256 × 256.
☐ (e) Only CT images can be windowed.

47. The matrix size of an image:

☐ (a) Is the same as the pixel size
☐ (b) Determines the number of pixels that are acquired
☐ (c) Determines the spatial resolution if the field of view is fixed
☐ (d) Applies only in diffusion-weighted imaging
☐ (e) Is the same as the field of view

48. When calculating the SNR in an image:

☐ (a) The noise is the standard deviation in the area of interest.
☐ (b) The noise is the standard deviation in a noisy background area.
☐ (c) The noise is the mean signal in a noisy background area.
☐ (d) The signal is the mean signal in the area of interest.
☐ (e) The signal is the maximum signal in the area of interest.

49. In the diagram showing an anatomical plane in Figure 11.6, the following statements are true:

☐ (a) The plane shown is a transverse plane.
☐ (b) The plane shown is a sagittal plane.
☐ (c) The plane shown is a coronal plane.
☐ (d) The plane shown is an axial plane.
☐ (e) The plane shown is a longitudinal plane.

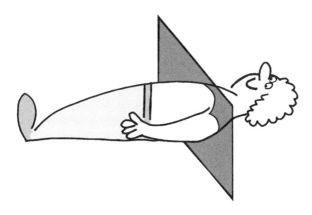

FIGURE 11.6
An anatomical plane for multiple-choice questions 49 and 50.

50. In the diagram showing an anatomical plane in Figure 11.6, the following statements are true:
    - ☐ (a) The plane shown is called a transverse plane.
    - ☐ (b) The plane shown is at right angles to a sagittal plane.
    - ☐ (c) The plane shown is a sagittal plane.
    - ☐ (d) The plane shown is at right angles to a longitudinal plane.
    - ☐ (e) The plane shown is at right angles to a transverse plane.

51. § In the diagram showing an anatomical plane in Figure 11.7, the following statement is true:
    - ☐ (a) The plane shown is a transverse plane.
    - ☐ (b) The plane shown is a sagittal plane.
    - ☐ (c) The plane shown is a coronal plane.
    - ☐ (d) The plane shown is an axial plane.
    - ☐ (e) The plane shown is a longitudinal plane.

FIGURE 11.7
An anatomical plane for multiple-choice question 51.

52. § At 1.5 T, the resonant frequency of protons is approximately:
   - ☐ (a) 64 kHz
   - ☐ (b) 42 MHz
   - ☐ (c) 64 MHz
   - ☐ (d) 85 MHz
   - ☐ (e) 85 kHz

53. § At 1.0 T, the resonant frequency of protons is approximately:
   - ☐ (a) 42 Hz
   - ☐ (b) 64 MHz
   - ☐ (c) 21 MHz
   - ☐ (d) 42 T
   - ☐ (e) 42 MHz

54. § If the gyromagnetic ratio is 42.56 MHz T^{-1}, then the Larmor frequency 85.12 MHz corresponds with a magnetic field of:
   - ☐ (a) 1.5 T
   - ☐ (b) 2.0 T
   - ☐ (c) 0.5 T
   - ☐ (d) 42.56 T
   - ☐ (e) 42.56 MHz

55. If a higher magnetic field strength is used:
   - ☐ (a) The precessional frequency of a given nucleus increases.
   - ☐ (b) The precessional frequency of a given nucleus decreases.
   - ☐ (c) The precessional frequency of a given nucleus might increase or decrease; it depends on what type of magnet is being used.
   - ☐ (d) There is no effect on the precessional frequency of a given nucleus.
   - ☐ (e) There is no effect on the gyromagnetic ratio of a given nucleus.

56. § Which one of the following expressions correctly describes the relationship of the resonant frequency of a nucleus to the magnetic field strength?
   - ☐ (a) Resonant frequency is directly proportional to field strength.
   - ☐ (b) Resonant frequency is inversely proportional to field strength.
   - ☐ (c) Resonant frequency is directly proportional to the square of the field strength.

□ (d) Resonant frequency is directly proportional to the square root of the field strength.

□ (e) Resonant frequency is directly proportional to the exponential of the field strength.

57. Select the correct statements:

□ (a) If the magnetic field strength is decreased, the precessional frequency of a given nucleus decreases.

□ (b) Resonant frequency and magnetic field strength are linked by an exponential relationship.

□ (c) If the magnetic field strength is decreased, the precessional frequency of a given nucleus increases.

□ (d) The gyromagnetic ratio is constant for a given nucleus.

□ (e) If the magnetic field strength is decreased, the gyromagnetic ratio for a given nucleus increases.

58. The rotating frame of reference:

□ (a) May also be called the laboratory frame of reference

□ (b) Rotates at a higher frequency than the Larmor frequency

□ (c) Rotates at a lower frequency than the Larmor frequency

□ (d) Rotates at the Larmor frequency

□ (e) Simplifies visualization of precessional motion

59. § When viewed in a rotating frame of reference, a vector precessing at a higher frequency than the Larmor frequency:

□ (a) Falls behind a vector precessing at the Larmor frequency

□ (b) Is stationary

□ (c) Draws ahead of a vector precessing at the Larmor frequency

□ (d) Is always 90° behind a vector precessing at the Larmor frequency

□ (e) Is always 90° ahead of a vector precessing at the Larmor frequency

60. Which of the following statements are true?

□ (a) The rotating frame should be viewed from an external stationary position.

□ (b) The rotating frame should be viewed from within the rotating frame of reference.

□ (c) The rotating frame should be viewed from an external position that is rotating at a different frequency from the rotating frame of reference.

☐ (d) The rotating frame of reference rotates around the y axis.

☐ (e) The rotating frame of reference rotates around the z axis.

61. A radio frequency coil can detect a signal from magnetization that is:

   ☐ (a) Parallel to the main magnetic field

   ☐ (b) Perpendicular to the main magnetic field

   ☐ (c) At right angles to the main magnetic field

   ☐ (d) At a 180° angle to the main magnetic field

   ☐ (e) At any angle to the main magnetic field

62. A superconducting magnet in an MRI system:

   ☐ (a) Is used to acquire the signal

   ☐ (b) Ensures that hydrogen nuclei are precessing at a known rate

   ☐ (c) Is usually found in low-field MR systems

   ☐ (d) Is usually found in high-field MR systems

   ☐ (e) Provides the main static field B_0

63. Which of the following are magnetic fields necessary for MRI?

   ☐ (a) Main field

   ☐ (b) Pilot field

   ☐ (c) Radio frequency field

   ☐ (d) Gradient fields for x, y, and z directions

   ☐ (e) Gradient field in z direction only

64. T1 values:

   ☐ (a) Do not depend on the strength of the main magnetic field

   ☐ (b) Are higher in high magnetic fields than in low magnetic fields

   ☐ (c) Are lower in high magnetic fields than in low magnetic fields

   ☐ (d) Do not depend on the strength of the radio frequency (RF) pulse

   ☐ (e) Are higher for stronger RF pulses

65. T2 values:

   ☐ (a) Are strongly dependent on the strength of the main magnetic field

   ☐ (b) Depend much less strongly on the strength of the main magnetic field than T1 values

- ☐ (c) Do not depend on the strength of the RF pulse
- ☐ (d) Are higher for stronger RF pulses
- ☐ (e) Are lower for stronger RF pulses

66. At a higher magnetic field strength, the value of T1 for a tissue is:
    - ☐ (a) Higher than T1 at a lower field strength
    - ☐ (b) About the same as T1 at a lower field strength
    - ☐ (c) Lower than T1 at a lower field strength
    - ☐ (d) Lower than T2 at the higher field strength
    - ☐ (e) Higher than T2 at the higher field strength

67. Which of the following are alternative terms for T1 relaxation?
    - ☐ (a) Spin-lattice relaxation
    - ☐ (b) Transverse relaxation
    - ☐ (c) Spin-spin relaxation
    - ☐ (d) Longitudinal relaxation
    - ☐ (e) Spin-echo relaxation

68. Which of the following are alternative terms for transverse relaxation?
    - ☐ (a) Spin-lattice relaxation
    - ☐ (b) T1 relaxation
    - ☐ (c) Spin-spin relaxation
    - ☐ (d) Longitudinal relaxation
    - ☐ (e) T2 relaxation

69. Which of the following are alternative terms for longitudinal relaxation?
    - ☐ (a) T1* relaxation
    - ☐ (b) T1 relaxation
    - ☐ (c) Transverse relaxation
    - ☐ (d) Spin-lattice relaxation
    - ☐ (e) T2* relaxation

70. § The recovery of longitudinal magnetization is directly proportional to:
    - ☐ (a) $\exp(-t/T2)$
    - ☐ (b) $1 - \exp(-t/T2)$
    - ☐ (c) $\exp(t/T1)$
    - ☐ (d) $1 - \exp(-t/T1)$
    - ☐ (e) $1 - \exp(-t/T2^*)$

71. § The decay of transverse magnetization is directly proportional to:
    - ☐ (a) $\exp(-t/T2)$
    - ☐ (b) $1 - \exp(t/T2)$
    - ☐ (c) $\exp(-t/T1)$
    - ☐ (d) $1 - \exp(t/T1)$
    - ☐ (e) $1 - \exp(t/T2^*)$

72. § The apparent decay of transverse magnetization in the presence of magnetic field inhomogeneities is directly proportional to:
    - ☐ (a) $\exp(-t/T2^*)$
    - ☐ (b) $\exp(-t/T1)$
    - ☐ (c) $1 - \exp(t/T2)$
    - ☐ (d) $\exp(t/T1^*)$
    - ☐ (e) $1 - \exp(t/T2^*)$

73. T1 is a time constant that represents:
    - ☐ (a) The time taken for the transverse magnetization to recover 63% of its maximum value
    - ☐ (b) The time taken for the longitudinal magnetization to recover 63% of its maximum value
    - ☐ (c) The time taken for the transverse magnetization to decay to 63% of its maximum value
    - ☐ (d) The time taken for the longitudinal magnetization to decay to 37% of its maximum value
    - ☐ (e) A value that is unique to each tissue

74. T2 relaxation:
    - ☐ (a) Arises from random interactions between the spins of neighboring nuclei
    - ☐ (b) Has an associated time constant, T2, which is the time taken for the transverse magnetization to decay to 37% of its maximum value
    - ☐ (c) Has an associated time constant, T2, which is the time taken for the transverse magnetization to lose 63% of its maximum value
    - ☐ (d) Is generally unrelated to the field strength
    - ☐ (e) Arises from interaction between the spins of hydrogen nuclei and the surrounding tissue

75. The return of longitudinal magnetization to equilibrium in a magnetic field:
    - ☐ (a) Is called T2 relaxation
    - ☐ (b) Is called spin-spin relaxation

☐ (c) Is called spin-lattice relaxation

☐ (d) Is a process of saturation recovery

☐ (e) Results in an unsaturated system

76. § A system is described as being saturated when:

☐ (a) A strong external field is applied but no RF energy has been supplied.

☐ (b) A strong external field is applied and RF energy has been supplied to tip the net magnetization vector through 180°.

☐ (c) No strong external field is present.

☐ (d) A strong external field is applied but the supply of RF energy means that there is no net magnetization in the direction of the external field.

☐ (e) A strong external field is applied and sufficient time has passed following a 90° RF pulse for the net magnetization vector to align with the external field.

77. T2* decay:

☐ (a) Is a longitudinal dephasing process

☐ (b) Is caused partly by susceptibility-induced field distortions from tissue

☐ (c) Arises from dephasing caused by inhomogeneities in the main field and also from intrinsic dephasing

☐ (d) Is a more rapid decay process than T2 decay

☐ (e) Takes more time than T2 decay

78. T2* decay:

☐ (a) Arises only from inhomogeneities in the main field

☐ (b) Arises from both random and fixed effects

☐ (c) Takes a shorter time than T2 decay

☐ (d) Is a dephasing process in the transverse plane

☐ (e) Represents recovery of longitudinal magnetization

79. Which of the following statements are correct?

☐ (a) T2* > T2 > T1

☐ (b) T1 > T2*

☐ (c) T1 > T2

☐ (d) T1 > T2 > T2*

☐ (e) T1 < T2

80. Which of the following statements are correct?

☐ (a) T1 < T2

☐ (b) T2 < T2*

    ☐ (c) T1 < T2*

    ☐ (d) T2 < T1

    ☐ (e) T2* < T2

81. § Which of the following statements is correct?

    ☐ (a) T2* < T1 < T2

    ☐ (b) T2* > T1 > T2

    ☐ (c) T2 < T2* < T1

    ☐ (d) T2 < T1 < T2*

    ☐ (e) T2* < T2 < T1

82. Which of the following statements are true?

    ☐ (a) Solid materials usually have a short T2.

    ☐ (b) Pure water has a short T2.

    ☐ (c) Solid materials and water usually have a long T1.

    ☐ (d) Fat has a short T1.

    ☐ (e) Fat has a short T2.

83. The T1 of pure water is approximately:

    ☐ (a) 20–30 ms

    ☐ (b) 200–300 ms

    ☐ (c) 2–3 seconds

    ☐ (d) 20–30 seconds

    ☐ (e) 2,000–3,000 ms

84. Which of the following statements are true?

    ☐ (a) Proteins and lipids have a short T1.

    ☐ (b) Pure water has a long T2.

    ☐ (c) Fat has a long T1.

    ☐ (d) Tissues with high water content have long T1 values.

    ☐ (e) Tissues with high macromolecular content have long T1 values.

85. A gradient coil in an MR system is a component included:

    ☐ (a) To improve the homogeneity of the magnetic field

    ☐ (b) To cause the strength of the magnetic field to vary linearly with the distance from the center of the magnet

    ☐ (c) To cause the strength of the magnetic field to vary with the square of the distance from the center of the magnet

    ☐ (d) To act as an RF receiver

    ☐ (e) To help localize the source of a signal

86. Gradient coils are used in MRI:
    - ☐ (a) To excite the nuclei
    - ☐ (b) For slice selection
    - ☐ (c) For signal localization
    - ☐ (d) For phase encoding
    - ☐ (e) For shielding

87. § Gradient coils in MRI:
    - ☐ (a) Are used only for slice selection
    - ☐ (b) Are used only for frequency encoding
    - ☐ (c) Are used only for phase encoding
    - ☐ (d) Are used for both frequency and phase encoding
    - ☐ (e) Are used for slice selection, frequency encoding, and phase encoding

88. To select a sagittal slice:
    - ☐ (a) The direction of the gradient magnetic field must be in the anterior-posterior direction.
    - ☐ (b) The gradient needs to change the value of the magnetic field from one side of the subject to the other.
    - ☐ (c) The gradient needs to change the value of the magnetic field from the anterior of the subject to the posterior.
    - ☐ (d) The direction of the gradient should be perpendicular to the plane of the sagittal slice.
    - ☐ (e) The direction of the gradient magnetic field must be from one side of the subject to the other.

89. To select a transverse slice:
    - ☐ (a) The gradient should change the value of the magnetic field in the subject in the cranio-caudal direction.
    - ☐ (b) The direction of the gradient magnetic field must be the same as the main field direction.
    - ☐ (c) The direction of the gradient should be perpendicular to the plane of the transverse slice.
    - ☐ (d) The gradient should change the value of the magnetic field in the subject in the left-right direction.
    - ☐ (e) The gradient should change the value of the magnetic field in the subject in the anterior-posterior direction.

90. § Why are slices in MRI often acquired with gaps between them?
    - ☐ (a) To reduce cross talk between adjacent slices
    - ☐ (b) To make the acquisition faster

☐ (c) To avoid the wraparound artifact

☐ (d) To prevent heating

☐ (e) To correct dephasing effects

91. The center frequency of an RF pulse is 63.84 MHz at the isocenter of the magnet. A slice is centered 20 mm from the isocenter, where the center frequency is 63.85 MHz. The gyromagnetic ratio is 42.56 MHz T^{-1}. The slice selection gradient used to achieve this is:

☐ (a) 0.01 MHz m^{-1}

☐ (b) 10 mT m^{-1}

☐ (c) 0.1175 mT m^{-1}

☐ (d) 0.5 MHz m^{-1}

☐ (e) 11.75 mT m^{-1}

92. The slice selection gradient strength is 10 mT m^{-1}. What transmitted bandwidth will provide a slice 10 mm thick? The gyromagnetic ratio is 42.56 MHz T^{-1}.

☐ (a) 0.1 mT

☐ (b) 4.256 × 10^{-3} Hz

☐ (c) 10 mT

☐ (d) 4.256 kHz

☐ (e) 2.35 MHz

93. In a system where the main field strength is 1 T, a slice of thickness 1 mm is centered 50 mm from the isocenter. The slice selection gradient field strength is 20 mT m^{-1} and the gyromagnetic ratio is 42.6 MHz T^{-1}. Which if the following statements are correct?

☐ (a) The transmitted bandwidth is 0.85 kHz.

☐ (b) The center frequency of the slice is 42.643 MHz.

☐ (c) The transmitted bandwidth is 0.02 mT.

☐ (d) The magnetic field strength at the center of the slice is 1.001 T.

☐ (e) A slice centered 25 mm from the isocenter could be obtained by doubling the gradient field strength.

94. § In a spin-echo sequence, the frequency encoding gradient is applied:

☐ (a) At the same time as the slice selection gradient

☐ (b) Immediately before the 90° RF pulse

☐ (c) Shortly before the signal measurement takes place

☐ (d) At the same time as the signal is measured

☐ (e) Immediately after the 90° RF pulse that causes saturation

95. An MR system with main field strength B_0 has a frequency encoding gradient with a receiver bandwidth of 2 mT. Which of the following are true?

☐ (a) Increasing the strength of the frequency encoding gradient increases the field of view.

☐ (b) Increasing the strength of the frequency encoding gradient decreases the field of view.

☐ (c) Increasing the strength of the frequency encoding gradient does not affect the field of view.

☐ (d) The range of field strengths when the gradient is applied runs from $(B_0 - 0.001)$ T to $(B_0 + 0.001)$ T.

☐ (e) The range of field strengths when the gradient is applied runs from $(B_0 - 0.002)$ T to $(B_0 + 0.002)$ T.

96. An MR system with main field strength B_0 has a frequency encoding gradient strength of 2 mT m^{-1}. Which of the following are true?

☐ (a) The range of field strengths when the gradient is applied depends on the value of the receiver bandwidth.

☐ (b) The range of field strengths when the gradient is applied runs from $(B_0 - 0.001)$ T to $(B_0 + 0.001)$ T.

☐ (c) Increasing the receiver bandwidth would increase the field of view.

☐ (d) Increasing the receiver bandwidth would decrease the field of view.

☐ (e) Increasing the receiver bandwidth would not affect the field of view.

97. § What is the field of view (FOV) in the frequency encoding direction if the frequency encoding gradient is 1 mT m^{-1}, the receiver bandwidth is 20 kHz, and the gyromagnetic ratio is 42.6 MHz T^{-1}?

☐ (a) 470 m

☐ (b) 0.47 m

☐ (c) 2.13 m

☐ (d) 0.213 m

☐ (e) 0.852 m

98. § What is the frequency encoding gradient strength if the FOV in the frequency encoding direction is 18 cm, the receiver bandwidth is 18 kHz, and the gyromagnetic ratio is 42.6 MHz T^{-1}?

☐ (a) 20 mT m^{-1}

☐ (b) 2 mT m^{-1}

☐ (c) 0.076 mT m^{-1}

☐ (d) 2 mT

☐ (e) 1 mT m^{-1}

99. § What is the receiver bandwidth if the frequency encoding gradient strength is 1.5 mT m^{-1}, the FOV in the frequency encoding direction is 45 cm, and the gyromagnetic ratio is 42.6 MHz T^{-1}?

☐ (a) 0.007 mT

☐ (b) 28.8 kHz

☐ (c) 28.8 MHz

☐ (d) 0.29 kHz

☐ (e) 2880 kHz

100. The phase encoding gradient is applied:

☐ (a) In the same direction as the frequency encoding gradient

☐ (b) Before the slice selection gradient

☐ (c) In the same direction as the slice selection gradient

☐ (d) At right angles to both the slice selection and frequency encoding gradients

☐ (e) Before the frequency encoding gradient

101. In a spin-echo sequence, a phase encoding gradient is applied in the *y* direction:

☐ (a) At the same time as the slice selection gradient

☐ (b) Once for every image

☐ (c) After slice selection but before signal measurement

☐ (d) At the same time as the signal is measured

☐ (e) Once for every row in the image

102. In phase encoding:

☐ (a) A small field of view is associated with large increments between successive gradients.

☐ (b) A small field of view is associated with small increments between successive gradients.

☐ (c) The size of the field of view is unaffected by the size of the increments between successive gradients.

☐ (d) At a given location in the phase encoding direction, the phase shift is incremented by the same amount for each successive gradient.

☐ (e) The phase difference between a pair of locations in the phase encoding direction is always the same.

103. § The minimum phase encoding gradient is -25 mT m^{-1} and the maximum phase encoding gradient is 25 mT m^{-1}. The increment in field strength between successive phase encoding gradients is 0.5 mT m^{-1}, and each gradient is applied for 1 ms. The gyromagnetic ratio is 42.6 MHz T^{-1}. The number of phase encoding steps is:

   ☐ (a) 99
   ☐ (b) 100
   ☐ (c) 101
   ☐ (d) 51
   ☐ (e) 199

104. § The minimum phase encoding gradient is -25 mT m^{-1} and the maximum phase encoding gradient is 25 mT m^{-1}. The increment in field strength between successive phase encoding gradients is 0.5 mT m^{-1}, and each gradient is applied for 1 ms. The gyromagnetic ratio is 42.6 MHz T^{-1}. The field of view is:

   ☐ (a) 47 cm
   ☐ (b) 4.7 cm
   ☐ (c) 21.3 cm
   ☐ (d) 2.13 cm
   ☐ (e) 11.7 cm

105. § The minimum phase encoding gradient is -25 mT m^{-1} and the maximum phase encoding gradient is 25 mT m^{-1}. The increment in field strength between successive phase encoding gradients is 0.5 mT m^{-1}, and each gradient is applied for 1 ms. The gyromagnetic ratio is 42.6 MHz T^{-1}. The pixel size in the phase encoding direction is:

   ☐ (a) 4.7 mm
   ☐ (b) 4.7 cm
   ☐ (c) 2.1 mm
   ☐ (d) 0.21 mm
   ☐ (e) 0.47 mm

106. Dephasing:

   ☐ (a) Arises from applying gradient fields
   ☐ (b) Arises from inhomogeneity of the main magnetic field
   ☐ (c) Takes place in the $x'z'$ plane and accounts for T1 relaxation
   ☐ (d) Takes place in the $x'y'$ plane and accounts for T2 relaxation
   ☐ (e) Arises from spin-spin interactions

107. Dephasing:
    ☐ (a) Is associated with T1 relaxation
    ☐ (b) Is associated with T2 relaxation
    ☐ (c) Is entirely a random process
    ☐ (d) Has both random and nonrandom components
    ☐ (e) Is entirely nonrandom

108. Which of the following statements about the spin-echo sequence and dephasing are true?
    ☐ (a) The spin-echo pulse sequence eliminates dephasing using a refocusing gradient pulse.
    ☐ (b) The spin-echo pulse sequence eliminates dephasing using a refocusing 180° RF pulse.
    ☐ (c) The spin-echo pulse sequence eliminates dephasing that has arisen from magnetic field inhomogeneities.
    ☐ (d) The spin-echo pulse sequence eliminates dephasing that has arisen from spin-spin interactions.
    ☐ (e) The spin-echo pulse sequence eliminates dephasing that has arisen from the use of gradients for spatial localization.

109. In a timing diagram for the spin-echo imaging pulse sequence:
    ☐ (a) The slice selection gradient is applied at the time of the 90° RF pulse.
    ☐ (b) The center of the frequency encoding gradient pulse is at time TE (time to echo) after the 90° RF pulse.
    ☐ (c) The time between the measurement of the signal and the first 90° RF pulse of the next repetition is (TR – TE).
    ☐ (d) The 180° RF pulse is applied at the same time as the frequency encoding gradient.
    ☐ (e) The signal is measured at a time TE/2 after the 180° RF pulse.

110. § Which of the following is the correct definition for TR in the spin-echo sequence?
    ☐ (a) TR is the time between the 90° RF pulse and the start of the phase encoding gradient.
    ☐ (b) TR is the time between the 90° RF pulse and the time at which the signal is measured.
    ☐ (c) TR is half the time between the 90° RF pulse and the time at which the signal is measured.

☐ (d) TR is the time period between successive 90° RF pulses.

☐ (e) TR is twice the time between the 90° RF pulse and the 180° refocusing RF pulse.

111. § Which of the following is the correct definition for TE in the spin-echo sequence?

☐ (a) TE is twice the time period between successive 90° RF pulses.

☐ (b) TE is the time period between successive 90° RF pulses.

☐ (c) TE is the time between the 90° RF pulse and the time at which the signal is measured.

☐ (d) TE is the time between the 90° RF pulse and the 180° refocusing RF pulse.

☐ (d) TE is the time between the 180° refocusing RF pulse and the following 90° RF pulse.

112. Tendons have a short T2 and long T1; in the spin-echo sequence they usually appear:

☐ (a) Bright in a long TE, long TR image

☐ (b) Dark in a long TE, long TR image

☐ (c) Bright in a short TE, short TR image

☐ (d) Dark in a short TE, short TR image

☐ (e) Bright in a short TE, long TR image

113. A lesion has a higher water content than the surrounding, otherwise similar tissue. In a spin-echo image the lesion:

☐ (a) Appears brighter than its surroundings in a long TE, long TR image

☐ (b) Appears darker than its surroundings in a long TE, long TR image

☐ (c) Appears brighter than its surroundings in a short TE, short TR image

☐ (d) Appears darker than its surroundings in a short TE, short TR image

☐ (e) Appears the same as its surroundings in a long TE, long TR image

114. Cerebrospinal fluid (CSF) has longer T1 and T2 values than white matter. Assuming that the proton density is the same for the two tissues, in a spin-echo image CSF appears:

☐ (a) Brighter than gray matter in a long TE, long TR image

☐ (b) Darker than gray matter in a long TE, long TR image

☐ (c) Brighter than gray matter in a short TE, short TR image

☐ (d) Brighter than gray matter in a long TE, short TR image

☐ (e) Darker than gray matter in a short TE, short TR image

115. § Arrange the T1-weighted images in Figure 11.8 in order of increasing slice thickness, starting with the thinnest. Apart from slice thickness, the acquisitions are otherwise identical.

☐ (a) Figures (c), (b), (a), (d)

☐ (b) Figures (a), (b), (d), (c)

☐ (c) Figures (d), (a), (b), (c)

☐ (d) Figures (a), (b), (d), (c)

☐ (e) Figures (d), (c), (a), (b)

116. § The slice thicknesses of the four T1-weighted images in Figure 11.8 are 1 mm, 2 mm, 5 mm, and 10 mm. Which image corresponds with the 10-mm slice thickness?

☐ (a) Figure (c)

☐ (b) Figure (a)

☐ (c) Figure (b)

☐ (d) Figure (d)

117. § You are told that the slice thickness of the image in Figure 11.8b is half the slice thickness for the image in Figure 11.8d. Both acquisitions had NEX = 1 and the same parameters other than slice thickness. If only NEX is changed and all the other acquisition parameters are kept the same, how many NEX will make the SNR for Figure 11.8b match that for Figure 11.8d?

☐ (a) 2

☐ (b) 4

☐ (c) $\sqrt{2}$

☐ (d) 16

☐ (e) $1/\sqrt{2}$

118. § Consider the T1-weighted images in Figure 11.9. Which image has the largest NEX? Apart from NEX, the acquisitions are identical.

☐ (a) Figure 11.9a

☐ (b) Figure 11.9b

☐ (c) Figure 11.9c

☐ (d) Figure 11.9d

119. § Which of the T1-weighted images in Figure 11.9 is associated with the smallest receiver bandwidth? Assume that the acquisitions are identical apart from the receiver bandwidth.

☐ (a) Figure 11.9a

☐ (b) Figure 11.9b

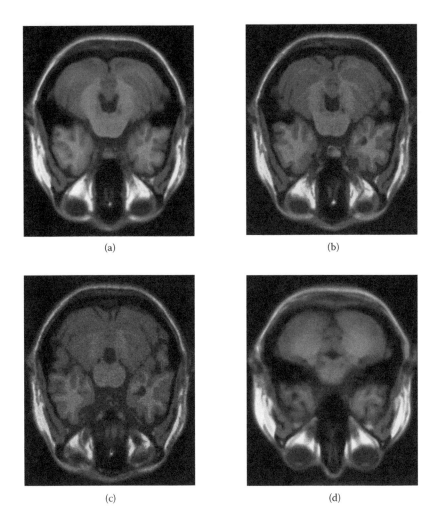

(a)   (b)

(c)   (d)

FIGURE 11.8
Four T1-weighted images, differing only in slice thickness, for multiple-choice questions 115 to 117. (Image data from the BrainWeb Simulated Brain Database, www.bic.mcgill.ca/brainweb/; Collins, D. L., et al., Design and construction of a realistic digital brain phantom, *IEEE Trans. Med. Imag.*, 17, 463, 1998. With permission.)

☐ (c) Figure 11.9c
☐ (d) Figure 11.9d

120. Which of the following statements concerning the T1-weighted images in Figure 11.9 are correct? The slice thickness for all the images is 2 mm, and only receiver bandwidth or NEX might differ.

☐ (a) The image with the highest SNR is (b).
☐ (b) The image with the highest SNR is (d).

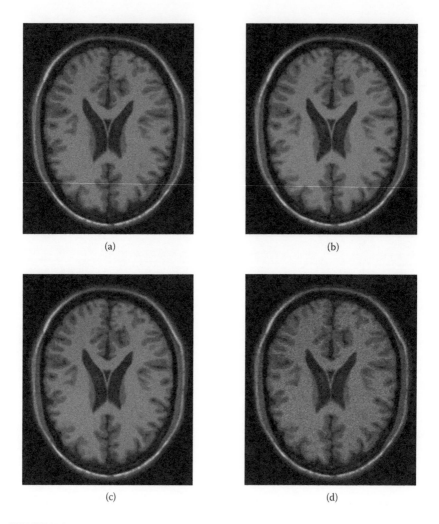

(a)   (b)

(c)   (d)

FIGURE 11.9
Four T1-weighted images, for multiple-choice questions 118 to 120. (Image data from the BrainWeb Simulated Brain Database, www.bic.mcgill.ca/brainweb/; Collins, D. L., et al., Design and construction of a realistic digital brain phantom, *IEEE Trans. Med. Imag.*, 17, 463, 1998. With permission.)

☐ (c) If the receiver bandwidth is the same for all four images, then NEX is smallest for image (b).

☐ (d) If NEX is the same for all four images, then the receiver bandwidth must be smallest for image (b).

☐ (e) If NEX is the same for all four images, then the receiver bandwidth must be smallest for image (d).

121. Increasing the bandwidth of the RF pulse:
    - ☐ (a) Increases the SNR in the image
    - ☐ (b) Decreases the SNR in the image
    - ☐ (c) Has no effect on the SNR in the image
    - ☐ (d) Increases the slice thickness
    - ☐ (e) Decreases the slice thickness

122. The SNR for an image can be increased by:
    - ☐ (a) Increasing the number of signals averaged
    - ☐ (b) Increasing the spatial resolution
    - ☐ (c) Increasing the FOV
    - ☐ (d) Increasing the slice thickness
    - ☐ (e) Increasing the number of phase encoding steps, but maintaining spatial resolution

123. Which of the following changes, made on its own without adjustments to other parameters, will lead to a reduction in the SNR of an image?
    - ☐ (a) The number of excitations is increased.
    - ☐ (b) Receiver bandwidth is increased.
    - ☐ (c) The number of phase encoding steps is reduced.
    - ☐ (d) The number of measurements made is reduced.
    - ☐ (e) The slice thickness is increased.

124. The total scan time for conventional spin-echo imaging is affected by:
    - ☐ (a) NEX
    - ☐ (b) TR
    - ☐ (c) TE
    - ☐ (d) Matrix size
    - ☐ (e) B_0

125. Which of the following could affect the scan time for a slice?
    - ☐ (a) Cardiac gating
    - ☐ (b) Bandwidth
    - ☐ (c) Slice thickness
    - ☐ (d) Matrix size
    - ☐ (e) Field of view

126. When imaging patients who are unable to keep still, it is advisable to shorten the acquisition time. Which of the following reduce the MRI acquisition time for a conventional spin-echo sequence?
    ☐ (a) Use fewer phase encoding steps.
    ☐ (b) Use fewer frequency encoding steps.
    ☐ (c) Decrease the number of acquisitions (NEX).
    ☐ (d) Reduce repetition time (TR).
    ☐ (e) Decrease the field of view (FOV).

127. Spatial resolution can be improved by:
    ☐ (a) Decreasing the slice thickness
    ☐ (b) Decreasing the matrix size
    ☐ (c) Decreasing the field of view (FOV)
    ☐ (d) Increasing the matrix size
    ☐ (e) Increasing the NEX

128. A 256 × 256 pixel MR image is acquired with a 2 mm slice thickness. The pixels are 1 mm in size in the frequency and phase encoding directions. Which of the following changes would increase the in-plane spatial resolution of the image? Assume that other parameters are held constant in each case.
    ☐ (a) Change to a 512 × 512 matrix.
    ☐ (b) Decrease to 1 mm slice thickness.
    ☐ (c) Increase the field of view in both frequency and phase encoding directions.
    ☐ (d) Change to a 128 × 128 matrix.
    ☐ (e) Decrease the field of view in both frequency and phase encoding directions.

129. Which of the following will reduce spatial resolution (i.e., make it worse)?
    ☐ (a) Decreasing the field of view and maintaining the matrix size
    ☐ (b) Maintaining the field of view and decreasing the matrix size
    ☐ (c) Increasing the field of view and maintaining the matrix size
    ☐ (d) Maintaining the field of view and increasing the matrix size
    ☐ (e) Decreasing the field of view and increasing the matrix size

130. § Which one of the following can be used as a fat suppression technique?
    ☐ (a) STIR
    ☐ (b) Gradient echo
    ☐ (c) FLASH
    ☐ (d) Spin-echo
    ☐ (e) Time of flight

131. § The T1 of a tissue is 1,000 ms. What value of TI will cause the signal from the tissue to be nulled in an inversion recovery sequence?
    ☐ (a) 500 ms
    ☐ (b) 1,000 ms
    ☐ (c) 693 ms
    ☐ (d) 370 ms
    ☐ (e) 630 ms

132. § In an inversion recovery sequence, TI is set to 300 ms. Which of the following tissues will have its signal nulled?
    ☐ (a) T1 = 1,040 ms, T2 = 208 ms
    ☐ (b) T1 = 2,165 ms, T2 = 433 ms
    ☐ (c) T1 = 433 ms, T2 = 108 ms
    ☐ (d) T1 = 208 ms, T2 = 52 ms
    ☐ (e) T1 = 600 ms, T2 = 120 ms

133. § It is suspected that lesions may be present that have longer T2 values than their surroundings. Which of the following sets of acquisition parameters for an inversion recovery sequence at 1.5 T will null the fat and also demonstrate the lesions if present?
    ☐ (a) TR = 2,000 ms, TI = 150 ms, TE = 15 ms
    ☐ (b) TR = 200 ms, TI = 150 ms, TE = 15 ms
    ☐ (c) TR = 2,500 ms, TI = 1,800 ms, TE = 10 ms
    ☐ (d) TR = 2,500 ms, TI = 150 ms, TE = 80 ms
    ☐ (e) TR = 3,000 ms, TI = 500 ms, TE = 75 ms

134. Which of the following statements are true regarding the conventional inversion recovery sequence?
    ☐ (a) Inversion recovery sequences give more heavily T1-weighted images than spin-echo sequences.
    ☐ (b) Inversion recovery sequences give more heavily T2-weighted images than spin-echo sequences.
    ☐ (c) TR should be long.

☐ (d) A longer TI will give proton density-weighted images instead of T1-weighted images.

☐ (e) A shorter TI will give proton density-weighted images instead of T1-weighted images.

135. Which of the following statements are true regarding the conventional inversion recovery sequence?

☐ (a) Acquisition times are long.

☐ (b) TI is the time between the 90° RF pulse and the following 180° RF pulse.

☐ (c) TI is the time between the 180° RF pulse and the following 90° RF pulse.

☐ (d) TR is short.

☐ (e) TE is short for T1-weighting.

136. Which of the following statements are true regarding the gradient-echo sequence?

☐ (a) Acquisition times are short compared with spin-echo imaging.

☐ (b) The signal-to-noise ratio is higher in a gradient-echo image than in the corresponding spin-echo image.

☐ (c) Pathology weighting replaces T2-weighting in gradient echo.

☐ (d) The 180° RF rephasing pulse is applied much sooner after the initial flip than in the spin-echo sequence.

☐ (e) There is no 180° RF rephasing pulse used in the gradient-echo sequence.

137. In the gradient-echo imaging pulse sequence:

☐ (a) The interval between successive α° RF pulses is called TE.

☐ (b) The interval between successive 180° RF pulses is called TR.

☐ (c) The interval between successive α° RF pulses is called TR.

☐ (d) There is no α° RF pulse.

☐ (e) There is no 180° RF pulse.

138. In the gradient-echo imaging pulse sequence:

☐ (a) The flip angle can be less than 90°.

☐ (b) Slice selection is performed using a gradient at the time of signal readout.

☐ (c) TR is longer than in spin-echo imaging to allow time for gradient rephasing.

☐ (d) The time between the flip angle and measurement of the signal is TE/2.

☐ (e) The time between the flip angle and measurement of the signal is TE.

139. In gradient-echo imaging, which of the following are true?

☐ (a) T2-weighted images cannot be obtained.

☐ (b) Proton density-weighted images cannot be obtained.

☐ (c) T2*-weighted images can be obtained.

☐ (d) T1*-weighted images can be obtained.

☐ (e) T1-weighted images can be obtained.

140. The Ernst angle in gradient-echo imaging:

☐ (a) Does not depend on the T2 of the tissue

☐ (b) Depends on the T1 of the tissue

☐ (c) Is equal to the flip angle used in the acquisition sequence

☐ (d) Depends on the TR of the acquisition sequence

☐ (e) Depends on the TE of the acquisition sequence

141. Cerebrospinal fluid (CSF) has longer T1 and T2 values than gray matter. Assuming that the proton density is the same for the two tissues, in a gradient-echo image gray matter appears:

☐ (a) Brighter than CSF in a long TE, short TR, small flip angle image

☐ (b) Darker than CSF in a long TE, short TR, small flip angle image

☐ (c) Brighter than CSF in a short TE, short TR, large flip angle image

☐ (d) Brighter than CSF in a short TE, short TR, small flip angle image

☐ (e) Darker than CSF in a short TE, short TR, large flip angle image

142. Motion artifacts in a two-dimensional sequence occur:

☐ (a) In the frequency encoding direction

☐ (b) In the phase encoding direction

☐ (c) In the slice selection direction

☐ (d) Owing to breathing

☐ (e) Owing to pulsating blood flow

143. Flowing blood:
    - ☐ (a) Can cause artifacts in an image even if it is outside the field of view
    - ☐ (b) Always leads to signal loss in a spin-echo image
    - ☐ (c) Can lead to signal loss or signal enhancement in a spin-echo image
    - ☐ (d) Causes signal loss in a spin-echo image if the flow is turbulent
    - ☐ (e) Has time-of-flight effects that always lead to signal enhancement in a gradient-echo image

144. Acquired images have motion artifacts that may be obscuring or mimicking a lesion. The operator decides to repeat the acquisition with the phase and frequency encoding directions swapped. For which types of movement is this strategy recommended?
    - ☐ (a) Changes in the subject's position
    - ☐ (b) Cardiac motion
    - ☐ (c) Blood flow
    - ☐ (d) Coughing
    - ☐ (e) Swallowing

145. Signal loss can occur in spin-echo imaging if material that has not been excited by the 90° RF pulse is present. Which of the following could be responsible for this sort of signal loss?
    - ☐ (a) Flowing blood
    - ☐ (b) Flowing cerebrospinal fluid
    - ☐ (c) Metal clips
    - ☐ (d) Air
    - ☐ (e) Tissue with short T1

146. Intravoxel phase dispersion:
    - ☐ (a) Arises when protons within a voxel are traveling with different velocities
    - ☐ (b) Arises when protons in neighboring voxels are traveling with different velocities
    - ☐ (c) Leads to signal enhancement near vessel walls
    - ☐ (d) Can be compensated for by using additional gradient fields
    - ☐ (e) Can lead to artifacts in the phase encoding direction

147. In gradient-echo imaging:
   - ☐ (a) Time-of-flight effects from flow always lead to signal enhancement.
   - ☐ (b) Time-of-flight effects from flow lead to greater signal enhancement when the flow is slow.
   - ☐ (c) Ghosting artifacts can be caused by pulsatile blood flow.
   - ☐ (d) Time-of-flight effects from flow can lead to signal loss or signal enhancement, depending on the flow velocity.
   - ☐ (e) High flow velocities result in washout signal loss.

148. Which of the following are methods of magnetic resonance angiography?
   - ☐ (a) Time of flight
   - ☐ (b) Phase contrast
   - ☐ (c) Echo planar
   - ☐ (d) Contrast enhanced
   - ☐ (e) Fat sat

149. Which of the following statements are true regarding magnetic resonance angiography (MRA)?
   - ☐ (a) Phase-contrast MRA involves use of an injected contrast agent.
   - ☐ (b) Time-of-flight MRA is performed using the spin-echo sequence.
   - ☐ (c) MIP is a way of acquiring images in the scanner.
   - ☐ (d) Phase-contrast MRA is directionally sensitive.
   - ☐ (e) Time-of-flight MRA can be performed using two- or three-dimensional sequences.

150. Which of the following statements are true regarding magnetic resonance angiography (MRA)?
   - ☐ (a) Contrast-enhanced MRA involves the injection of a bolus of contrast agent.
   - ☐ (b) Contrast-enhanced MRA involves the use of additional gradients.
   - ☐ (c) Time-of-flight techniques are more suitable than contrast-enhanced techniques for large field of view angiography.
   - ☐ (d) Time-of-flight techniques rely on the shortening of the T1 of blood by a contrast agent.
   - ☐ (e) There is no directional information in angiograms produced using contrast-enhanced MRA.

11.2   Answers to Multiple-Choice Questions

1. b	6. a
2. a	7. c, d
3. c	8. a, b, d
4. d	9. c, e
5. c	

FEEDBACK ON DEGREES AND RADIANS FOR QUESTIONS 7 TO 12

You should be able to use radians and degrees equally confidently for multiples of 90°. The key fact to memorize is that an angle of $\pi/2$ radians is the same angle as 90°. If you remember this, then you can deduce that 360° must be 2π radians, that 180° is π radians, and so on. Also, 1 radian must be $90/(\pi/2)$ degrees, or about 57°. The phase difference between the functions $\sin(x)$ and $\cos(x)$ is 90°. When $x = 0$, $\sin(x) = 0$ and $\cos(x) = 1$. When $x = 90°$, $\sin(x) = 1$ and $\cos(x) = 0$.

10. a, e	25. c, d, e
11. a, b, c, d, e	26. c, d
12. c, e	27. b, e
13. a, b, c, e	28. a, b, c
14. a, e	29. b
15. a, b, e	30. a
16. a, c, e	31. b, d, e
17. c, e	32. a
18. a, b, d	33. b, c, d
19. a, b, e	34. c, e
20. a, c	35. b, d
21. b, d	36. a, d
22. a, d	37. a, b, c, e
23. a, b, c	38. b, c
24. a, c	39. b, e

**FEEDBACK ON ENERGY AND THE ELECTROMAGNETIC
SPECTRUM FOR QUESTIONS 37 TO 39**

The RF pulses in MRI are electromagnetic radiation with frequencies in the radio frequency range. Radio frequency radiation has longer wavelengths and lower frequencies than all the other kinds of electromagnetic radiation. It was stated in Chapter 2 that radio frequency electromagnetic radiation has low energy, which is insufficient to cause ionization. The relationship between energy in joules (E) and wavelength (λ) is $E = hc/\lambda$, where h is Planck's constant and c is the speed of light.

40. a, b, e
41. a, b, d
42. b, c
43. c
44. a, c, d
45. a, b, c, d, e
46. a, c
47. b, c
48. b, d
49. a, d
50. a, b
51. c
52. c
53. e
54. b
55. a, e
56. a
57. a, d
58. d, e
59. c
60. b, e
61. b, c
62. b, d, e

63. a, c, d
64. b, d
65. b, c
66. a, e
67. a, d
68. c, e
69. b, d
70. d
71. a
72. a
73. b, e
74. a, b, c, d
75. c, d, e
76. d
77. b, c, d
78. b, c, d
79. b, c, d
80. d, e
81. e
82. a, c, d
83. c, e
84. a, b, d

FEEDBACK ON THE STATES OF WATER FOUND IN THE BODY FOR QUESTIONS 82 TO 84

Pure water has a long T1 and a long T2; for example, T1 may be as long as 3,000 ms. Note, however, that water in the body is not in the pure state, but may be bound, to proteins or other macromolecules for example. Fluids in the body may therefore exhibit shorter T1 and T2 times than pure water; for example, T1 is typically 400 to 800 ms.

85. b, e

86. b, c, d

87. e

88. b, d. (a) and (e) are only true if the subject was arranged so that the direction described was also the direction of the main field.

89. a, b, c

90. a

91. d, e

92. a, d

93. a, b, c, d, e

FEEDBACK ON UNITS FOR QUESTIONS 91 TO 93

In Questions 91 to 93 conversion from field strength in tesla to units of frequency, and from bandwidth in units of frequency to tesla, is made using the Larmor equation.

94. d

95. b, d

96. a, c. (b) is only true under particular circumstances that were not stated in the question.

97. b

98. b

99. b

100. d, e

101. c, e

102. a, d

103. c

104. b

105. e

106. a, b, d, e

107. b, d

108. b, c

109. a, b, c, e

110. d

111. c

112. b, d

FEEDBACK ON TENDONS AND
LIGAMENTS FOR QUESTION 112

The combination of long T1 and short T2 for the tendon means that when recovery curves are drawn, the tendon curve tends to be of small amplitude, whatever combination of TE and TR is chosen. However, tendons do sometimes appear bright in spin-echo images; this is an artifact called the *magic angle effect*. The effect is seen in tendons and ligaments that are oriented at an angle of about 55° to the main magnetic field. At this angle the dipolar interactions, which normally cause the short T2, become zero. This means that T2 increases by about a factor of 100, and a signal is visible for tendons.

113. a, d	120. a, d
114. a, d, e	121. b, d
115. a	122. a, c, d, e
116. d	123. b, c, d
117. b	124. a, b, d
118. b	125. a, d
119. b	

FEEDBACK ON CARDIAC GATING FOR QUESTION 125

The equation concerning scan time that applies to this question is Equation 9.4. Cardiac gating does not appear explicitly in the expression, but it is related to TR. If cardiac gating is in use and the heartbeat is irregular, the next RF pulse is not necessarily triggered during every cardiac cycle, as the system may wait for the next good heartbeat. This increases the overall scan time.

126. a, c, d	129. b, c
127. a, c, d	130. a
128. a, e	131. c

FEEDBACK ON SIGNAL NULLING
FOR QUESTIONS 130 TO 132

STIR stands for short TI inversion recovery. In this technique the choice of inversion time (TI) can be made to null the signal from tissue with a chosen T1. The nulled tissue is very often fat. To null a tissue with longitudinal relaxation time T1, the inversion time used is 0.693 T1. Other techniques that can be used to suppress the signal from fat include chemical presaturation. Spatial presaturation may be used to reduce artifacts from flowing blood.

132. c	139. a, c, e
133. d	140. a, b, d
134. a, c, d	141. b, c
135. a, c, e	142. b, d, e
136. a, e	143. a, c, d, e
137. c, e	144. b, c
138. a, e	

FEEDBACK ON MOTION ARTIFACTS
FOR QUESTIONS 142 TO 144

Motion artifacts occur in the phase encoding direction for two reasons. First, motion along any of the phase encoding gradients will result in the signal being mismapped because the phase accumulation is wrong. Second, motion generally takes place more slowly than the time that is used to perform frequency encoding, and so the motion will only become apparent because of the longer time that elapses between phase encoding steps. Both periodic movements and random movements cause artifacts of this kind. Periodic motion includes, for example, cardiac motion, blood flow, and CSF motion, while random motion includes breathing, changes in position, swallowing, and coughing. For periodic motion it can be useful to swap the phase and frequency encoding directions to help differentiate between a lesion and an artifact. The location of the artifact will change under these circumstances, but a lesion would stay in the same place.

145. a, b	148. a, b, d
146. a, d, e	149. d, e
147. a, c	150. a, e

Index

*For Product Safety Concerns and Information please contact
our EU representative GPSR@taylorandfrancis.com Taylor & Francis
Verlag GmbH, Kaufingerstraße 24, 80331 München, Germany*

T - #0018 - 160425 - C0 - 234/156/17 [19] - CB - 9781584889014 - Gloss Lamination